Amal Almousa

Production of citric acid by locally isolated Aspergillus niger using agriculture wastes

Amal Almousa

Production of citric acid by locally isolated Aspergillus niger using agriculture wastes

Investing agricultural waste by microorganisms to produce organic acids

Noor Publishing

Imprint
Any brand names and product names mentioned in this book are subject to trademark, brand or patent protection and are trademarks or registered trademarks of their respective holders. The use of brand names, product names, common names, trade names, product descriptions etc. even without a particular marking in this work is in no way to be construed to mean that such names may be regarded as unrestricted in respect of trademark and brand protection legislation and could thus be used by anyone.

Cover image: www.ingimage.com

Publisher:
Noor Publishing
is a trademark of
International Book Market Service Ltd., member of OmniScriptum Publishing Group
17 Meldrum Street, Beau Bassin 71504, Mauritius

Printed at: see last page
ISBN: 978-620-0-07480-5

Production of citric acid by locally isolated *Aspergillus niger* using agriculture wastes

Amal Ali Almousa

TABLE OF CONTENTS

Acknowledgment

First, thanks for God for helping me to do this work. I would like to express my deep gratitude and sincerely thanks to my Supervised Prof. Dr. Eman Hussein Ashour, Professor of Microbiology at Botany & Microbiology Department, College of Science, King Saud University for her kind supervising this work and continuous valuable guidance.

I would like also to acknowledge Dr Monera AL-Othman, Doctor of Microbiology at Botany & Microbiology Department, College of Science, King Saud University .

I cannot forget to thank full to Sisters Amani, Geyhan & Nadine for their assistance.

Acknowledgment

First, thanks for God for helping me to do this work. I would like to express my deep gratitude and sincerely thanks to my Supervised Prof. Dr. Eman Hussein Ashour, Professor of Microbiology at Botany & Microbiology Department, College of Science, King Saud University for her kind supervising this work and continuous valuable guidance.

I would like also to acknowledge Dr Monera AL-Othman, Doctor of Microbiology at Botany & Microbiology Department, College of Science, King Saud University .

I cannot forget to thank full to Sisters Amani, Geyhan & Nadine for their assistance.

LIST OF FIGURES

LIST OF ABBREVIATIONS

ANOV	Analysis Of Variance
ATP	Adenosine Triphosphate
CHNS	Carbon,Hydrogen,Nitrogen and Sulfur
DNA	Deoxyribonucleic Acid
EGR	Rawabi garden east of Riyadh
EMCC132	Reference strain
FAR	Farm in Arka
FDA	Food and Drug Administration
FDD	A Farm in the dam Diriya
FGD	A Farm in AL-Ghad
FDR	A Farm in Dharme
FDRA	A Farm in Dirab (A)
FDRB	A Farm in Dirab (B)
FHA	Farm in Hayer
FKH	A Farm in AL-Kharj
FMN	Farm in Mansuriya
FMZ	A Farm in AL-Muzahmiyya
FQWA	Farm way Qassim (A)
FQWB	Farm way Qassim (B)
FTHA	Farm in Thumama (A)
FTHB	Farm in Thumama (B)
GRAS	Generally Regarded As Safe
HPLC	High-Performance Liquid Chromatography Technique

KSUD	King Saud University Diriya‹ Garden
KSUM	King Saud University Malaz‹ Garden
LSD	Least Significant Difference
NAD	Nicotin amid adenine dinucleotide
NTG	N-methyl-N-nitro-N-nitrosoguandine
NWH	North Wadi Hanifa
PDA	Potato Dextrose Agar
PH	potential of Hydrogen
RPM	Revolutions Per Minute
SE	Standard Error
SPR	Slam Park in the center of Riyadh
SWH	South Wadi Hanifa
TCA	Tri-Carboxylic acid
UV	Ultraviolet

ABSTRACT

Citric acid is one of the most common products which have a never ending demand in the global market. Citric acid fermentation is one of the primitive fermentations but still its production is increasing with passage of time. So, the goal of the research focused on 1) isolation and screening of different local strains of *A. niger* for their ability to grow and produce citric acid; 2) selection of the most potent isolates for citric acid production; 3) optimization certain environmental and nutritional parameters affecting citric acid production by the most efficient fungal strains; 4) utilization of certain available and cheap agricultural wastes as alternatives to citric acid chemical substrates and 5) studying of the influence of mutagenic agents (physical and chemical mutagens) on the most potent *A. niger* strains.

From twenty soil samples collected, several *Aspergilli* strains were isolated and purified. Based on morphological and microscopically characterization, ninety six isolates were identified as *Aspergillus niger*.

All *A. niger* isolates understudying were varied in their capability to produce citric acid on indicator plats medium. Through this an easily and quick technique, six *A. niger* isolates were selected which gave highest productivity of citric acid after 24 hours incubation.

Based on the productivity of citric acid in shake flask cultures after 3 days, the highest two local *A. niger* FQW and KSUD, were selected for further experiments with the reference strain, *A. niger* EMCC132.

Results of the time course analysis of tested fungal strains indicated that optimal time incubation for the maximum citric acid production reached statistically to maximum values on 5th day of fermentation.

Through series of experiments were designed on various physico-chemical fermentation parameters to establish the optimal conditions for citric acid overproduction, it could be summarized as: optimum temperature, 28°C; initial pH 6.5; inoculums size $3.0X10^6$ spores/ml; shaking culture (100rpm); maltose as a carbon source 40 g/L; peptone as a nitrogen source 3 g/L; Sodium phosphate as phosphorus source; 2% of methanol as enhancer.

Through these environmental and nutritional parameters, citric acid production was clearly improved and gave 14-12 times higher that reached to 133.31 and 109.29 mg/10ml comparing to the original yields 9.07 and 9.98 mg/10ml by *Aspergillus niger* FQW and KSUD, respectively.

When local agricultural wastes and byproducts (pineapple peel, sugarcane bagasse, potato peels; sugarcane molasses and dates molasses) utilized for citric acid production, local strain *A. niger* FQW produced highest amount of citric acid, reached up to 256.94 mg/10 ml when grew on sugarcane molasses as a sole carbon source.

In addition, local strain *A. niger* KSUD produced highest amount of citric acid, reached up to 223.77 mg/10 ml when grew on pineapple peels as a sole carbon and nitrogen source.

By assessing the obtained findings among used raw materials, it is considering the fact that for production of commercially valuable citric acid by tested *A. niger* strains, molasses (either sugarcane or date) and pineapple peels proved to be appropriate substrates from the cost point of view, it is highly possible to consider these effective processes as beneficial, economical or cost-effective methods.

In a trial for improvement of citric acid production by using physical and chemical mutagens, it found that *A. niger* FQW gave highest amount of citric acid and researched to 41.98 mg/L after exposed to UV at 254 nm for 180 min comparing with its parental productivity which gave 13.918 mg/L. In addition, muted strain of *A. niger* FQW demonstrated more stable and affected at a concentration of MTG, 750 milligrams per liter for 60 min which gave citric acid more twofold higher (31.3 mg /L) than its parental activity (13.9 mg /L).

1. INTRODUCTION

Citric acid (2-hydroxy-propane-1,2,3-tricarboxylic acid) derives its name from the Latin word citrus, the citrus tree. It is ubiquitous in nature and exists as an intermediate in citric acid cycle when carbohydrates are oxidized to carbon dioxide. Citric acid was first produced commercially in England around 1826 from imported Italian lemons (**Rojoka *et al.*, 1998**).

Citric acid is non toxic and easily oxidized in the human body. Because of its high solubility, palatability and low toxicity, it can be used in food, biochemical and pharmaceutical industries. These uses have placed greater stress on increased citric acid production and search for more efficient fermentation process. The world wide demand of citric acid about 6.0×10^5 tons per year and is bound to increases day by day (**Ali *et al.*, 2001**).

In 2007, the world wide production of citric acid was approximately 1.6 million tons according to the business communications Company (BCC)'s recent studies of fermentation (**Pandey *et al.*, 2013**). Moreover due to its large application and low price, the citric acid consumption is expected to grow significantly until 2009, and this raises the need for industries to search for new technological alternative and for cost reduction in citric acid production (**Vandenberghe *et al.*, 2004**). Citric acid has a variety of applications, 70% of it used in food and beverage industries, 12% in the pharmaceutical industry and 18% in other industries (**Pandey *et al.*, 2001; Soccol *et al.*, 2003**).

A potentially promising economical and efficient source for producing citric acid is to use the fungus *A. niger* due to its higher capacity to accumulate acid when compared to other organisms (**Yokoya, 1992; Pazouki *et al.*, 2000 and Adham, 2002**). Although many microorganisms can be used to produce citric acid, *A. niger* remains the main industrial producer. Specific strains that are capable of overproducing citric acid have been developed for various types of fermentation processes (**El-Holi and Al-Delaimy, 2003**).

Citric acid production by fermentation is the most economical and widely used way of obtaining the product (**Priede and Latvian 2005**). There are basically three different types of fermentation process; these are solid state fermentation, the liquid surface culture and the submerged fermentation process (**Adham, 2002**). Nowadays, more than 90 % of the citric acid produced in the world is obtained by fermentation, which has its own advantages. The submerged technique is widely used for citric acid production. It is estimated that about 80% of world production is obtained by submerged fermentation. This fermentation process is employed in large scale bases requires more sophisticated technique and rigorous control. On the other hand, it presents several advantages such as higher productivity and yields, lower labor costs, and lower contamination risk **(El-Holi and Al-Delaimy, 2003; Barrington and Kim 2008).**

The supply of natural citric acid is limited and the demand can only be satisfied by biotechnological fermentation processes. Now, *Aspergillus niger* is exclusively used for industrial scale production of citric acid **(Suzuki *et al.*, 1996).** The optimization of fermentation conditions are of primary importance in the development of any fermentation process owing to their impact on the economy and practicability of the process **(Hang and Woodams 1985).**

Moreover, possibilities have been done for using some by-products and agroindustrial residues in citric acid production, on the other hand, solving in serious environmental problems **(Soccol, 1996; Pintado *et al.*,1998; Soccol and Vandenberghe, 2003; Darani and Zoghi, 2008 and Rodrigues *et al.*, 2010).** Citric acid production by *A. niger* depends on several specific nutritional and environmental conditions as well as the particular strain of microorganisms.

Mutations are abrupt and so are the hereditary modifications in the genetic material. The organism containing the DNA are not static molecules and their bases are frequently exposed to natural or artificial agents can cause modification in their structure or in chemical composition

(**Zaha, 2003 and Griffiths** *et al.,* **2006**). The increase in citric acid productivity has been achieved using mutation and strain selection. Mutation strains with certain characteristics such as enhanced citric acid production and increased fermentation rate have been previously selected after submitting the genetic material to physical or chemical mutagenic agents (**Rohr** *et al.,***1983, Ikram-Ul Hag** *et al.,***2001; Griffiths** *et al.,***2006, and Lotfy** *et al.,* **2007b).**

Therefore, strategy of this research focuses on a detailed study of citric acid production. This goal achieved through the following aspects:

1. Isolation and screening of different local strains of *A. niger* for their ability to grow and produce citric acid.

2. Selection of the most potent isolates for citric acid production.

3. Optimization certain environmental and physiological parameters affecting citric acid production by the most efficient fungal strains.

4. Utilization of certain available and cheap agricultural wastes as alternatives to citric acid chemical substrates

5. Studying of the influence of mutagenic agents (physical and chemical mutagens) on the most potent *A. niger* strains.

2. REVIEW OF LITERATURE

2.1 Citric Acid

Citric acid derives its name from the Latin word *citrus*, the citrus tree. It was first isolated from lemon juice in 1784 by a Swedish chemist Carl Scheele. Citric acid is a nearly universal intermediate product of metabolism and its traces are found in virtually all plants and animals. Citric acid ($C_6H_8O_7$) is a tricarboxylic acid with a molecular weight of 210.14 Da. Its IUPAC name is 2-hydroxy-propane-1,2,3-tricarboxylic acid (Fig. 1). At room temperature, citric acid is a white hygroscopic crystalline powder, readily soluble in water. It is solid at room temperature, melts at 153°C and when heated above 175°C, it decomposes through the loss of carbon dioxide and water (**Kubicek and Rohr 1986**).

Fig. (2.1): Chemical structure of Citric Acid (a hydroxyl polycarboxylic acid)

2.1.1 Sources of Citric Acid

Citric acid exists in a variety of fruits and vegetables, most notably citrus fruits. **Penniston *et al.*, (2008)** reported Lemons and limes have particularly high concentrations of the acid; it can constitute as much as 8% of the dry weight of these fruits (about 47 g/L in the juices). The concentrations of citric acid in citrus fruits range from 0.005 mol/L for oranges and grapefruits to 0.30 mol/L in lemons and limes. Within species, these values vary depending on the cultivar and the circumstances in which the fruit was grown.

Citric acid was first produced commercially in England around 1826 from imported Italian lemons. As its commercial importance increased, Italian lemon grows started producing it to establish a virtual monopoly during the most of the nineteenth century. Lemon juice remained the commercial source of citric acid until 1919 when the first industrial process using *Aspergillus niger* began in Belgium. Citric acid had been synthesized from glycerol by **Grimoux and Adams (1880)** and later from symmetrical dichloroacetone. Several other synthetic routes using different starting materials have since been published, but chemical methods have so far proved uncompetitive with fermentation mainly because the starting materials are worth more than the final product.

Among the organic acids industrially produced, citric acid is the most important in quantitative terms with an estimated annual production of about 1.4 million tons. The annual growth of its demand/consumption rate is around 3.5-4.0 % (**Soccol *et al.,* 2006**).

2.1.2 Citric Acid Benefits

Organic acids have long history of being utilized as food additives and preservatives for preventing food deterioration and extending the shelf life of perishable food ingredient.

Citric acid is widely used to impart a pleasant, tart flavour to foods and beverages. It also finds applications as a function of additive detergents, pharmaceuticals, cosmetics and toiletries. About 64 % of U.S. citric acid usage in 2004 was for foods and beverages, 22 % for detergents and cleaning products and 10 % for pharmaceutical and nutritional products. About 2 % went into cosmetics and toiletries. Around 2 % were used in different applications. The actual price of citric acid is about $1 to $1.3 per kilo. Due to the numerous applications and low prices of citric acid, consumption is expected to grow strongly, and considering slight price increases until 2009, the market value for citric acid will exceed $2 billion (**Partos, 2005**).

2.2. Microbial Production of Citric Acid

2.2.1 History of Citric Acid Production

Historically, citric acid was first extracted from lemon juice, explaining the origin of its Latin name *citrus*. Citric acid was first isolated from lemon juice in (1784) by the Carl Wilhelm Scheele, a Swedish chemist by a culture of *Penicillium glaucum* on sugar medium. After a few years, he isolated two new fungal strains with the ability to accumulate citric acid, which were designated *Citromyces* (*Penicillium*). In 1917, James Currie, an American, discovered that citric acid could be produced industrially by *Aspergillus niger*. He found that numerous strains of *Aspergillus niger* produced significant amounts of citric acid. The most important finding was that *A. niger* grew well at pH values around 2.5–3.5 and high concentrations of sugars favour citric acid production. After that, citric acid can be produced by various types of microorganisms. Among yeasts, several species have been evaluated for this purpose, such as *Candida, Torula, Saccharomyces* and *Pichia*. Although the specie *Candida* is used for the commercial production of citric acid, it also produces substantial amounts of unwanted isocitric acid (as reported by **Soccol *et al.,* 2006**).

It is known that several yeasts produce citric acid, especially species belonging to the genera *Candida, Hansenula, Pichia, Debaromyces, Torula, Torulopsis, Kloekera, Saccharomyces, Zygosaccharomyces* and *Yarrowia*. Among yeast species: *Saccaromicopsis lipolytica*, *Candida tropicalis*, *C. oleophils, C. guilliermondi, C. parapsilosis* and *C. citroformans* were detected as a good citric acid producing candidates (**Kapelli *et al.*, 1978; Rane and Sims 1993; Wojtatowics *et al.*, 1993**). While today, this production is not economical. As a disadvantage, the fermentation by yeasts led to the formation of large quantities of isocitric acid as an unwanted byproduct.

2.2.2.Citric Acid-Producing Filamentous Fungi

The problem in the production of citric acid for yeasts is the simultaneous formation of isocitrate. The main advantages of using *Aspergillus niger* are its ease of handling, its ability to ferment a variety of cheap raw materials and high yields. The yield of citric acid from these strains often exceeds 70 % of the theoretical yield on the carbon source (**Papagianni 2007**).

Wehmer (**1893**) was the first to demonstrate that *Citromyces* (now *Penicillium*) accumulated citric acid in a medium containing sugar and inorganic salts. Since then, many organisms have been found to accumulate citric acid: *A. niger*, *Aspergillus awamori*, *Aspergillus nidulans*, *Aspergillus fonsecaeus*, *Aspergillus luchensis*, *Aspergillus phoenicus*, *Aspergillus wentii*, *Aspergillus saitoi*, *Aspergillus flavus*, *Absidia* sp., *Acremonium* sp., *Botrytis* sp., *Eupenicillium* sp., *Mucor piriformis*, *Penicillium janthinellum*, *Penicillium restrictum*, *Talaromyces* sp., *Trichoderma viride* and *Ustulina vulgaris* (**Papagianni 2007**).

In 1917 **Currie** found that some strains of *A. niger* were able to grow in a medium containing sugars and salts at an initial pH of 2.5-3.5. Throughout their growth, these strains excreted large amounts of citric acid, which established the basis for industrial production.

Although many microorganisms can be employed to produce citric acid, *A. niger* is still the main industrial producer. In fact, specific strains that are able to overproduce citric acid in different types of fermentation processes have been developed.

2.2.3. Growth of the fungus *Aspergillus niger*

Fungi generally exhibit four basic growth stages: the spore which is a dormant phase, followed by the spore germination or lag phase, the growth or hyphae phase, and the spore formation phase. During the growth phase, the fungus develops tubular filaments known as hyphae, and these hyphae branch out repeatedly to form a mass know as the mycelium. During this phase, the rate of production of biomass is exponential (Fig. 2.2). Citric acid production is known to occur when fungal growth is limited (**Vaija and Linko, 1986**).

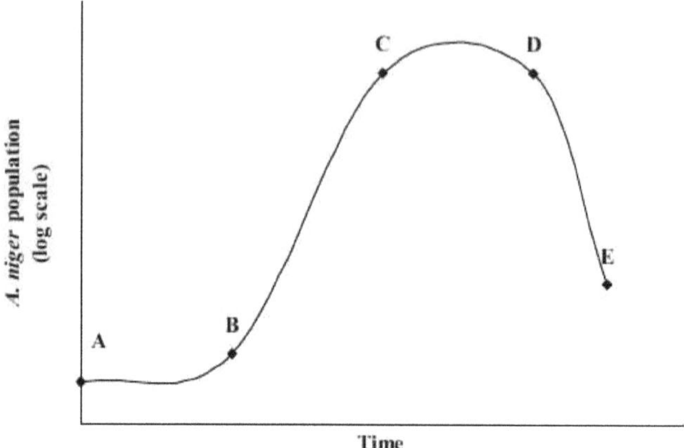

Fig. (2.2): Growth stages of *A. niger*: A to B initial lag phase; B to C exponential growth phase; C to D Stationary phase; and D to E decline phase.

New organisms originate from spores produced asexually by a mature *A. niger* mycelium. To reproduce, fungi develop special hyphae called sporangiophores which stand erect and develop numerous spores at their tip. These spores are released into the environment and will remain dormant until proper conditions allow for their development (**Nason,**

1968). The metabolism of spores will generally be activated by physical or chemical factors present in the environment, such as moisture, light, temperature, and substrates such as carbohydrates.

The introduction of any microorganisms, including *A. niger*, to a new environment requires the synthesis of enzymes and intermediates to initiate nutrient ingestion for growth. The duration of the lag phase therefore depends on several factors, including environmental conditions and level of inoculum. For *A. niger* grown under submerged conditions, high levels of sugar have resulted in an extended lag phase, because such conditions repress the formation of the enzyme α-keto-glutarate dehydrogenase, an enzyme involved in the citric acid cycle (**Papagianni, 2004; Hossain *et al.,* 1984**).

When nutrients become limited, especially for the older hyphae, a stationary phase is exhibited where growth is virtually stopped. According to **Papagianni (2004)** this phase establishes the balance between hyphal mass increase and decrease. If nutrients are lacking over a prolonged period, the fungal hyphae undergoing a stationary growth phase may experience a loss of cell mass and even die as food reserves are deoleted. Death of fungal hyphae also occurs in the presence of toxins or unfavorable environmental conditions such as intolerable pH, and high levels of heavy metals and salts.

2.2.4. Biochemistry of Citric Acid Production by *Aspergillus niger*

Citric acid or citrate, the ionic form of citric acid, is a major intermediate compound produced by the tri-carboxylic acid (TCA) cycle. A product resulting from the oxidation of glucose, namely glycolysis, pyruvate ($HO(CO)_2CH_3$) is the compound required to initiate the TCA cycle (Fig. 2.3). Pyruvate must first be converted to acetaldehyde (CH3COH) and then Acetyl CoA (Fig. 2.3) to activate the TCA. The first intermediate compound produced by the TCA cycle is citrate ($C_3H_5O(COO)_3^{-3}$) which is the ionic form of citric acid ($CH_2HOC(COOH)_3CH_2$). In *A. niger*, the TCA cycle occurs in the matrix of the mitochondrion (a structure generating ATP and controlling cell growth). Citric acid (2-hydroxy-1,2,3-propanetricarboxylic acid) is mainly produced by fermentation using *Aspergillus niger*. Citric acid accumulation by *A. niger* is achieved by two main metabolic pathways: (1) the catabolic pathway where hexose is transformed into pyruvate and acetyl-Coenzyme A (Acetyl-CoA) by glycolysis, and; (2) citric acid formation by the TCA cycle (**Alvares-Vasquez *et al.*, 2000**). Once formed, citrate or citric acid can be excreted out of the cell, if conditions prohibit the second stage of the TCA cycle or can continues within the cycle to finally generate Oxaloacetic acid (Oxaloacetate) and several compounds used to store energy such as ATP and NADPH (Fig. 2.3).

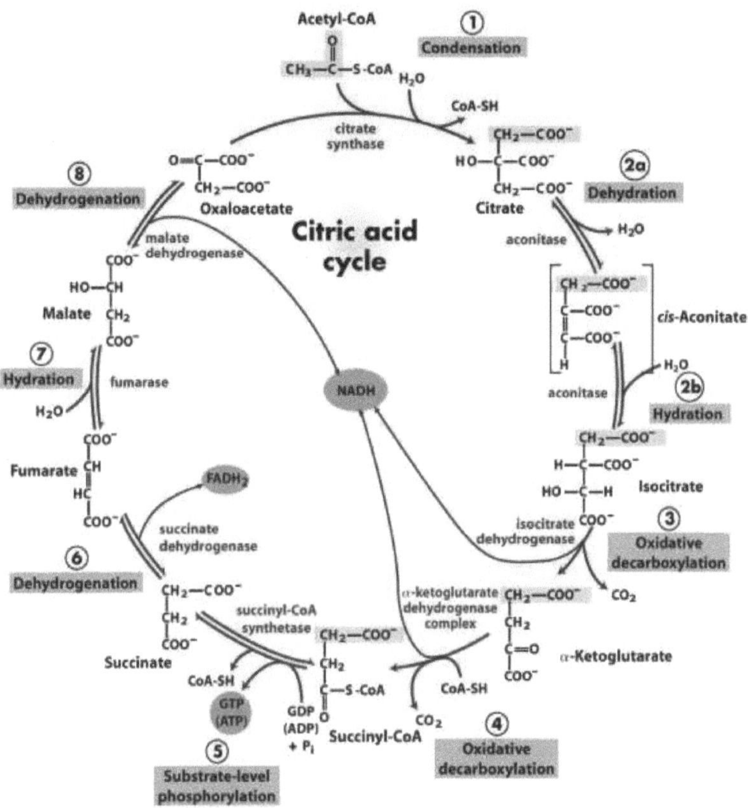

Fig. (2.3): The Tri-carboxylic acid (TCA) cycle responsible for the production of citrate (the ionic form of citric acid). Citrate can be used as a major intermediate in the TCA cycle, or can be excreted outside the cell wall. If nutrients become limited, the citrate accumulated outside the cell wall can be reabsorbed and degraded within the TCA cycle by the enzyme aconitase with the eventual generation of Oxaloacetate and several compounds used to store energy such as ATP and NADPH. (https://celestemohan.wordpress.com/2013/04/10/tca-the-citric-acid-cycle/).

Several theories were proposed to explain the extracellular excretion of citric acid by *A. niger*. Citric acid excretion is said to drop the general pH of the environment in the vicinity of the hyphae to enable the fungus to biologically compete against its enemies and to enhance the mobility of compounds such as heavy metals and for phosphate (**Sayer and Gadd, 2001**).

According to **Anastassiadis *et al.*, (2005)** the intracellular accumulation of citric acid results from the inactivation of citrate degrading enzymes such as aconitase or isocitrate dehydrogenases. **Ahmed *et al.*, (1972)** mentioned that the inactivation of these enzymes results from a low internal level of free amino acids. Nevertheless, a significant increase in cell protein level has been observed during citric acid accumulation (**Jernejc *et al.*, 1992**). Accordingly, the TCA cycle produces major compounds required for biomass formation. The mitochondrial citrate carriers of *A. niger* accumulate citrate but the phenomenon still need more investigation (**Papagianni, 2007**).

Citric acid is an intermediate product of the Tri-carboxylic acid (TCA) cycle. This cycle is initiated in the presence of pyruvate produced from the glycolysis or oxidation of sugars such as sucrose, glucose and fructose. Within the fungal cell, the TCA cycle is the main process generating ATP (Adenosine Tri-phosphate) used to transport energy, and NAD

(Nicotinamide adenine dinucleotide) an enzyme used in redox reactions such as the oxidation of sugars, and in the transport of electrons (**Moore-Landecker 1996**).

Within the TCA cycle, citric acid is an intermediate metabolic compound unless accumulated and excreted outside the fungal cell wall (**Papagianni 2007**), when the TCA cycle is inhibited by: a lack of enzymes resulting from a shortage of nutrients such as nitrogen; by enzyme inactivation as a result of low pH conditions, and; by insufficient nutrient for growth in the presence of an abundant sugar supply.

Jernecj *et al.,* (**1992**) observed that optimal citric acid production occurred when the fungal mycelium stopped accumulating lipids and nucleic acid, resulting in large changes in intra- and extra-cellular proteins and amino acids, and intra-cellular TCA cycle acids.

Citric acid overproduction requires a unique combination of several unusual nutrient conditions, i.e. excessive concentrations of carbon source, hydrogen ions, and dissolved oxygen, and suboptimal concentrations of certain trace metals and phosphate, that synergistically influence the yield of citric acid (**Kristiansen and Sinclair 1978**).

2.3.Factors affecting Citric Acid Production

Aspergillus niger has been used commercially for the first time in 1923 for citric acid production and it remained the organism of choice for commercial production because it produces more citric acid per time unit. Recently a wide range of citric acid production has been reported in response to different levels of nutrient supplementation (**Barrington and Kim 2008; Bari** *et al.,* **2009**).

Citric acid productivity can be increased by optimizing its fermentation variables: fermentation techniques, Incubation time, Fermentation temperature, inoculums size, Initial pH, Stirrer speed incubation (agitation), carbon source, nitrogen source, additives and Nutrients' concentration.

2.3.1. Fermentation techniques

Fermentation of a substrate to citric acid is directly related to the quality and quantity of the sugar source. The nature of the substrate which in turn depends on the carbon source will also have a marked influence on the metabolic activity of the microbial strains. Citric acid production synthesis by fermentation is the most economical and widely used way of obtaining this product. More than 90 % of the citric acid produced in the world is obtained by fermentation, which has its own advantages: operations are simple and stable, the plant is generally less complicated and needs less sophisticated control systems, technical skills required are lower, energy consumption is lower and frequent power failures do not critically affect the functioning of the plant (**Soccol** *et al.,* **2006**).

The industrial citric acid production can be carried in three different ways: by submerged fermentation, surface fermentation and solid-state fermentation process (**Yokoya 1992**).

2.3.1.1. Submerged fermentation

The submerged technique is widely used for citric acid production. It is estimated that about 80 % of world production is obtained by submerged fermentation (**Pallares** *et al.,* **1996; Vandenberghe** *et al.,* **1999**). It presents several advantages such as higher productivity and yields, lower labour costs, lower contamination risk and labour consumption. Submerged fermentation can be carried out in batch, fed batch or continuous systems, although the batch mode is more frequently used.

The commercial production of citric acid by filamentous fungi such as *A. niger* is mainly conducted using submerged fermentation techniques in which the growing solution contains sugars such as glucose or sucrose (**Leangon *et al.*, 2000**).

Submerged fermentation can be controlled to provide an optimal environment for the growth of filamentous and non-filamentous fungi. In this type of fermentation process, sterility of material can be controlled by maintaining the proper temperature, pH, dissolved oxygen, and mixing conditions such as speed of mixing, size of vessel, and fermentation volume. In the absence of mixing (static culture) for submerged fermentation, filamentous fungi form a mat of hyphae at the liquid surface (**Kumar *et al.*, 2003**).

When mixing is present, the culture is generally submerged in the solution and *A. niger* tends to develop pellets of hyphae, unless the mixing speed is too high and shears the hyphae away from each other. During the fermentation process, excessive mixing leads to cell structure damage, morphological changes, variations in growth rate and less efficient product formation (**Papagianni, 2004**).

2.3.1.2. Solid state fermentation

Solid state fermentation, also known by »Koji« process, was first developed in Japan where abundant raw materials such as fruit wastes and mainly rice bran are available. It is the simplest method for citric acid production and it has been an alternative method for using agro-industrial residues. Solid-state culture is characterized by the development of microorganisms in a low-water activity environment on an insoluble material that acts both as physical support and source of nutrients (**Xu *et al.*, 1989; Pallares *et al.*, 1996; Vandenberghe 2000**).

The solid culture process is completed within 96 hours under optimal conditions (**Pandey *et al.*, 2001**). The most common organism used in solid-state fermentation is *A. niger*. However, there have also been reports with yeasts. The strains with large requirements of nitrogen and phosphorus are not ideal microorganisms for solid culture due to lower diffusion rate of nutrients and metabolites occurring at lower water activity in solid-state process. The presence of trace elements may not affect citric acid production so harmfully as it does in submerged fermentation, thus, substrate pretreatment is not required. This is one of the important advantages of the solid culture (**Hossain *et al.*, 1984**).

Citric acid production can be done by solid-state fermentation or with submerged fermentation. To date, most industrial processes are carried out with submerged fermentation of *Aspergillus niger* (**Adham 2002**).

2.3.2. Agitation

In submerged fermentations, agitation is necessary for good mass and heat transfer rates. Mechanical forces resulting from the rotation of impellers can affect filamentous micro-organisms in various ways, causing rate variations of growth and product formation, development of different morphological forms and breakage of filaments. Since filamentous micro-organisms are of great industrial importance, the relationship between mechanical parameters, morphology and productivity has attracted the attention of many investigators (**Papagianni *et al.*, 1994**).

Particular morphological forms, often agitation induced, have been considered as more suitable for certain fermentations. Several authors have noted shorter and wider cells in fungal ®laments, with greater branching frequencies and reduced aggregate sizes under conditions of high agitation (**Van Suijdam *et al.*, 1980; Ujcova *et al.*, 1980**).

However, optimum stirrer speeds have been reported for many processes, while, side reactions and unwanted morphologies characterized stirrer speeds out of the ranges of interest(**Ujcova *et al.*, 1980**). Studying the behavior of three citric acid producing *Aspergillus niger* strains, **Ujcova *et al.*, (1980)** observed that higher speeds resulted in thicker, densely branched filaments, while production was optimum within a narrow range of speeds and a drop in productivity resulted from higher stirrer speeds.

Much effort has focused on the discovery of new metabolites and on improvement of the respective production methods. Information is, therefore, needed about the fundamental growth behavior with respect to morphology. The characterization of the morphology of filamentous micro-organisms started many years ago. Early attempts to link mechanical parameters with morphology and production, were limiting by the lack of the means to quantify morphology (**Paul *et al., * 1999**).

The impeller tip speed, the circulation frequency around the impeller, the power dissipation rate, are examples of parameters which been correlated with production rates and morphology parameters, as in the works of **Van Suijdam *et al., * (1980)**, and **Smith *et al., * (1990)**. However, when the validity of such correlations was tested under different operation scales, it was either met with apparent success or it was found difficult to prove the generality of the proposed relationships (**Smith *et al., * 1990**).

2.3.3. Incubation temperature

Most fungi produced a higher yield of citric acid at temperatures ranging between 25 and 35°C. Higher temperatures lead to enzyme denaturation and moisture loss while lower temperatures lead to reduced metabolic activity (**Adinarayana and Prabhakar 2003**).

Fawole and Odunfa (2003) reported that a temperature of 30°C was optimum for metabolite production by *A. niger*. The fungal cells showed signs of adverse growth and metabolic production if an ideal temperature was not provided during fermentation.

In solid state fermentation *A. niger* grows under conditions which are closer to its natural environment. Considering this important fact, *A. niger* may be more capable of producing certain enzymes and metabolites which usually are produced at low yield in submerged cultures (**Ellaiah** *et al.,* **2003**).

2.3.4. Initial pH Value

The pH of a culture may change in response to microbial metabolic activities. The most obvious reason is the secretion of organic acids, such as citric acid, which will cause the pH decrease. Changes in pH kinetics also depend highly on the microorganism. With *Aspergillus* sp., *Penicillium* sp. and *Rhizopus* sp., pH can drop very quickly to less than 3.0. For other groups of fungi such as *Trichoderma, Sporotrichum, Pleurotus* sp., pH is more stable between 4.0 and 5.0. The nature of the substrate and production technique also influence pH kinetics (**Yokoya 1992**). In this way initial pH must be very well defined and optimized depending on the microorganism, substrate and production technique.

All fermentations are started from spores even if the culture is "scaled up" through seed fermenters before being transferred to production fermenters. Secondly, the pH for citric acid production needs to be low (pH≤2). A low pH reduces the risk of contamination of the fermentation with other microorganisms. A low pH also inhibits the production of unwanted organic acids (gluconic acid, oxalic acid) and this makes the recovery of citric acid from the broth simpler. The uptake of ammonia by the germinating spores causes a corresponding release of protons that lower the pH to approximately the right level after spore germination has occurred. Increasing the pH to 4.5 during the production phase reduces the final yield of citric acid by up to 80% (**Papagianni *et al.*, 1999**).

Some investigators have reported that specific fungi can grow at a pH below 2 to compete against bacteria (**Fawole and Odunfa, 2003**). A low pH inhibits the production of unwanted organic acids such as gluconic acid and oxalic acid. Increasing the pH to 4.5 during the production phase reduces the final yield of citric acid by up to 80% (**Papagianni *et al.*, 1999**).

2.3.5. Carbon Source

Filamentous fungi are heterotrophic organisms, meaning that they require as energy and carbon source, organic compounds such as glucose, fructose and sucrose. Other compounds such as polysaccharides, amino acids, lipids, organic acids, proteins, alcohols and hydrocarbons can also be used by fungi (**Mattey, 1992; Krishna, 2005**). The preferred carbon source for *A. niger* is sugar because this substrate can be oxidized directly into pyruvate which is responsible for initiating the TCA cycle. The sugar was reported for citric acid production by *A. niger* include maltose, sucrose, glucose, mannose and fructose. Sucrose is preferred over other sugars, because *A. niger* produces an extracellular mycelium bound invertase that is active at low pH and rapidly hydrolyzes sucrose (**Kubicek and Rohr 1989**).

Certain raw materials, grape pomace, banana extract, sugar cane baggasse, sugar beet molasses, and apple peels have been used to produce citric acid using *A. niger* because of their high sugar content (**Wang, 1998; Wayman and Mattey 2000**). These wastes also provide essential micronutrients.

During glycolysis, an excess of glucose is known to lead to the over-production of citric acid (**Leangon *et al.*, 2000**), while a low glucose level may lead to the accumulation of oxalic acid. In the mitochondrion, citrate results from the condensation of oxaloacetate with the acetyl CoA enzyme. If the TCA cycle cannot be completed because of a lack of enzymes, citrate is excreted, and then oxaloacetate is produced from pyruvate as a result of pyruvate carboxylase. Also, limitation of carbon can cause cell aging and autolysis in fungal culture, which leads to fragmentation and heavily vacuolated hyphae (**Papagianni, 2004**) which ultimately affects citric acid production.

The carbon source for the citric acid fermentation has been the focus of much study, frequently with a view to the utilizing polysaccharide sources. In general, only sugars that are rapidly taken up by the fungus allow a high final yield of citric acid. Polysaccharides, unless hydrolyzed, are generally not a useful raw material for citric acid fermentation because they are broken down too slowly to match the high rate of sugar catabolism required for citric acid production (**Mattey, 1992**).

The slow hydrolysis of polysaccharides is due to the low activity of the hydrolytic enzymes at the low pH that is necessary for producing citric acid. According to **Kubicek and Röhr (1989)** sucrose is preferable to glucose as *A. niger* has a potent extracellular mycelium bound invertase that is active at low pH and rapidly hydrolyzes sucrose. Superiority of sucrose over glucose and fructose was demonstrated by **Gupta *et al.*, (1976)**, **Hossain *et al.*, (1984)** and **Xu *et al.*, (1989)**.

2.3.6. Nitrogen Source

Citric acid production is directly influenced by the concentration and nature of the nitrogen source. Physiologically, ammonium salts are preferred, such as urea, ammonium nitrate and sulphate, peptone, malt extract, *etc.* (**Grewal *et al.,* 1995; Vandenberghe *et al.,* 1999**). An advantage of using ammonium salts is that the pH declines as the salts are consumed and a low pH is a requirement of citric acid fermentation (**Mattey, 1992**).

Complex media such as molasses are rich in nitrogen-containing compounds and rarely need to be supplemented with a nitrogen source. The high purity media that are used mainly in research laboratories are generally supplemented with ammonium salts, particularly ammonium nitrate and ammonium sulfate, to provide the necessary nitrogen. Other sources of nitrogen that have been used include urea and yeast/malt extract (**Xu *et al.,* 1989**). Reports concerning the effects of nitrogen and phosphate limitations have been contradictory. According to **Shu and Johnson (1948)** phosphate does not need to be limiting for citric acid production; however, when trace metal levels are not limiting, additional phosphate results in side reactions and increased biomass growth (**Papagianni *et al.,* 2006**). **Kubicek and Röhr (1989)** showed that citric acid accumulated whenever phosphate was limited even when nitrogen was not. In contrast,

Kristiansen and Sinclair (1978) using continuous culture concluded that nitrogen limitation was essential for citric acid production.

In the nutrient medium for solid substrate fermentation, nitrogen is generally supplied as nitrate, ammonia or amino acids (**Krishna, 2005**). The most suitable sources of nitrogen are yeast extracts, peptone, ammonium sulphate, ammonium chloride and ammonium nitrate. However, it is necessary to maintain pH values in the first day of fermentation prior to a certain quantity biomass production.

The concentration of nitrogen source required for citric acid fermentation is 0.1 to 0.4 g/L (**Kubicek and Röhr 1989**). High nitrogen concentrations increase fungal growth and sugar consumption but decrease the amount of citric acid produced (**Papagianni *et al.,* 2006**).

2.3.7. Phosphorous source

Phosphorous plays an important role in fungal morphology. It is the main component of ATP, a compounds used to accumulate and transport energy within cells. Phosphorous is also involved in cell multiplication and metabolite production. It may be supplied as KH_2PO_4 or K_2HPO_4 (**Abou-Zeid and Ashy, 1984**) and the range of 0.06 to 0.32 g/l was found to optimize citric acid yields (**Shu and Johnson, 1984**). The excess concentration of exogenous phosphorous can induce low yield of citric acid.

In the first works it was verified that the presence of phosphate in the medium had a great effect on the yield of citric acid. Low levels of phosphate have positive effect on citric acid production. This effect acts at the level of enzyme activity and not at the level of gene expression. On the other hand, the presence of excess of phosphate leads to a decrease in the fixation of carbon dioxide, which in turn increases the formation of certain sugar acids, and the stimulation of growth (**Kubicek C.P. and Rohr 1989; Grewal and Kalra 1995; Vandenberghe** *et al.,* **1999**).

2.4. Raw materials usage for fungal citric acid production

Molasses for long considered as a waste product of the sugar industry is now termed as a by e product due to its low price compared to other sugar sources and the presence of minerals, organic and inorganic compounds. Molasses is used in the production of alcohol, organic acid and single cell proteins. The organic and inorganic components present in molasses may inhibit the fermentation process and hence need to be treated by suitable methods to make it suitable for citric acid production. Some of the commonly followed procedures include treatment with ferrocyanide (**Baby and Saad 1996**), sulphuric acid, tricalcium phosphate, tricalcium phosphate with HCl, tricalcium phosphate with HCl followed by sephadex fractionation (**Panda *et al.*, 1984**). Molasses became more efficient by treatment with ammonium oxalate followed by treatment with diammonium phosphate. Molasses treated by this method was found to serve as a better substrate in producing citric acid compared to other methods commonly practiced (**Angumeenal & Venkappayya, 2005**). The above molasses medium on supplementation with selective metal ions as stimulants made citric acid fermentation more successful. Agroeindustrial wastes are frequently used as substrates in fermentation as an inexpensive source. Apple pomace (**Hang & Woodams 1984**), carob pod (**Roukas, 1998**), carrot waste (**Garg & Hang 1995**), coffee husk (**Shankaran & Lonsane, 1994**), corn cobs (**Hang &Woodams 1998**), grape pomace (**Hang &.Woodams 1985**), kiwi fruit peel (**Hang &. Woodams, 1987**), kumara (**Lu, Brooks &. Maddox, 1997**), orange waste (**Aravantinos-Zafiris, Tzia, Orea poulou, & Thomopoulos, 1994**), date syrup (**Roukas & Kotzekidou 1997; Moataza, 2006**), pineapple waste (**Tran l.i.Sly 1998**), banana extract (**Sassi, Ruggeri, Specchia, & Gianetto, 1991**) potato chips waste (**Sanat, & Korish, 2007**), and pumpkin were tried successfully as substrates for citric acid formation. Pumpkin, either as a single or a mixed substrate with molasses is known to produce good quantities of citric acid (**Majumdar, Khalid, Munshi, Alam, & Or-Rashidet, 2010**). A waste from jackfruit was found to be a good and economical substrate for citric acid fermentation. Artrcocarpus heterophyllus (Jack fruit) is a large tree grown in tropical countries and is one of the common fruits in South India. The fruiting perianths (bulbs), seeds and rind constitute 29, 12 and 59% of the ripe fruit, respectively. The rind portion includes the carpel fibre which holds the fruity portion intact. Chemical analysis of this carpel fibre indicates the presence of sugars and minerals and hencewas used as a substrate for citric acid production. It was treated chemically to extract the reducing sugar present. Batch fermentation using A. niger was followed and the results indicate that jackfruit carpel fibre can serve as a substrate for citric acid production (**Angumeenal & Venkappayya, 2005a, 2005b, 2005c**). When this substrate is completely analysed and the limiting substances identified, steps can be taken to remove them and make it a more efficient substrate for citric acid fermentation. Tuber crops belonging to the family Araceae namely Colocassia antiquorum, Aponogetannatans and Amorphophallus campanulatus are cultivated in large quantities for their edible portion. These tubers were chemically treated and utilized as substrates for citrate production by fermentation using *A. niger* by batch fermentation. Their fermentation capabilities were improved by adding trace elements (cadmium, molybdenum, chromium and lead) in optimum quantities (**Angumeenal *et al.,* 2003a, 2003b**). While using *A. campanulatus* as a substrate along with citric acid, succinic acid was also produced in high amounts. In fact, succinic acid produced was higher than citric acid. This is due to the increased activity of aconitase in the later stages of fermentation. Hence this substrate can be further explored for succinic acid production using some growth promoters. The potential of *A. campanulatus* in producing citric acid was enhanced by the addition of metal ions.

Table (2.1): Comparison of citric acid production between results in the current research and the literature values.

Organism	Substrate	Fermentation configuration	Citric acid yield (g/L)	Reference
Aspergillus niger	*Parkia biglobosa* fruit pulp	Shake flask	1.15	Present study
Aspergillus niger	Coconut husk	Shake flask	1.00	Sukesh *et al.*, (2013)
Aspergillus niger	Whey/sucrose	Shake flask	106.50	El-Holi and Al-Delaimy (2003)
Aspergillus niger 14/20	Pumpkin	Shake flask	10.35	Majumder *et al.*, (2009)
Aspergillus niger	Tapioca	Shake flask	1.60	Sukesh *et al.*, (2013)
Aspergillus niger	Apple	Shake flask	2.10	Sukesh *et al.*, (2013)
Aspergillus niger 14/20	Molasses	Shake flask	7.72	Majumder *et al.*, (2009)
Aspergillus niger	Bagasse	Shake flask	0.24	Vaishnavi *et al.*, (2012)
Aspergillus niger	Beet molasses	Shake flask	240.10	Lotfy *et al.*, (2007a)
Aspergillus niger	Corn steep liquor	Shake flask	10.50	Lotfy *et al.*, (2007a)
Aspergillus niger ATTCC 9142	Beet molasses	Shake flask	35.00	Roukas (1991)

2.5. Tools for Citric Acid Production Improvement

The most employed technique to improve citric acid producing strains has been by inducing mutations in parental strains using mutagens. Among physical mutagens, g-radiation (**Bonatelli and Azevedo, 1983; Gunde-Cimerman, 1986 ;Islam *et al.,* 1986**) and UV-radiation (**Pelechova *et al.,* 1990**) have often used. To obtain hyperproducer strains, frequently UV treatment could be combined with some chemical mutagens, e.g. aziridine, N-nitroso-N-methylurea or ethyl methane-sulfonate (**Musilkova *et al.,* 1983**).

By using a suitable selection technique on model medium with non-specific carbon sources, a strain yielding high amounts of citric acid from unusual substrates can be obtained from the mutants produced.

Another approach for strain improvement has been the para-sexual cycle, as first described by **Pontecorvo *et al.,* (1953)**. According to **Das and Roy (1978)**, diploids displayed higher citric acid yields compared to their parent haploids, but tended to be less stable (**Bonatelli and Azevedo, 1983**). Protoplast fusion appeared to be a promising tool to extend the range of genetic manipulation of *A. niger* with respect to citric acid production. **Kirimura et al. (1988a)** studied protoplast fusion of production strains. They were able to obtain fusants with acid production capacities exceeding those of the parent strains in solid state fermentation, but not in submerged fermentation. Some other aspects of strain improvement could be the resistance to detrimental constituents of fermentation raw materials, capability of utilizing raw materials (starch, cellulose, pectin containing and other waste materials). However, there is no single effective technique to achieve hyper-producing mutants and much remains to be done in this area.

2.6. Applications of citric acid

The food industry consumes about 70 % of total citric acid produced and pharmaceutical industries consume about 12 %, and the remaining 18 % are consumed by other industries (**Pandey** *et al.,* **2001; Soccol** *et al.,* **2006**). Citric acid is used in the food, beverage, pharmaceutical, chemical, cosmetic and other industries for applications such as acidulation, antioxidation, flavour enhancement, preservation, plasticizer and as a synergistic agent. Citric acid is used to flavour the drinks, jams and jellies, candies, water ice and wines.

Table (2.2): Summarized of the most of the citric acid

Applications	Industry	Functions
Beverages	Wines and ciders	• Prevents browning in some white wines. Prevents turbidity of wines and ciders. Used in pH adjustment
	Soft drinks and syrups	• Provides tartness. Stimulates natural fruit flavour. As acidulant in carbonated and sucrose based beverages
Food	Jellies, jams and preservatives	• Used in pH adjustment. Acts as acidulant. Provides the desired degree of tartness, tang and flavour. Increases the effectiveness of antimicrobial preservatives
	Dairy products	• As emulsifier in ice creams and processed cheese. Acidifying agent and antioxidant in many cheese products.
	Candies	• Acts as acidulant. Provides tartness. Minimizes sucrose inversion. Produces dark colour in hard candies. Prevents crystallization of sucrose
	Frozen fruit	• Protects ascorbic acid by inactivating trace metals. Lowers pH to

		inactivate • oxidative enzymes.
	Fats and Oils	• Synergist for other antioxidants, as sequestrant. Stabilizing action
	Animal feed	• Feed complementation
Agriculture		• Micronutrient evaluation in fertilizers. • Enhances P availability in plants
Pharmaceutics	Pharmaceuticals	• As effervescent in powders and tablets in combination with bicarbonates. Anticoagulant. • Provides rapid dissolution of active ingredients. Acidulant in mildly astringent formulation
	Cosmetics and toiletries	• Buffering agent. pH adjustment. • Antioxidant as a metallic–ion chelator

Other	Industrial applications	<ul><li>Acts as buffer agent. Sequestring metal ions. Neutralizes bases.</li><li>Used in nontoxic, noncorrosive and biodegradable processes that meet current ecological and safety standards</li></ul>
	Metal cleaning	<ul><li>Removes metal oxides from the surface of ferrous and nonferrous metals, for operational cleaning of iron and copper oxides.</li><li>In electroplating, copper plating, metal cleaning, leather tanning, printing inks, bottle washing compounds, floor cement, textiles, photographic reagents, concrete, plaster, refractories and moulds, adhesives, paper, polymers, tobacco, waste treatment, chemical conditioner on teeth surface, ion complexation in ceramic manufacture</li></ul>

3. MATERIALS AND METHODS

3.1. Soil Samples Collection

Soil samples were collected from upper layer (up to 25 cm) of various locations in Riyadh region, Saudi Arabia (as mentioned in the maps, Fig. 7.1-3 in Appendix). Several diverse habitats in different areas were selected for the isolation of various local fungal strains. The soil samples were placed in plastic bags, closed tightly and stored in refrigerator.

3.2. Isolation and Purification of Fungal Strains

For fungi isolation from collected soil samples, ten gram of each soil sample was suspended in 90 ml of sterilized distilled water. Subsequently, fungal cultures were isolated by serial dilutions in sterilized saline and plated onto Petri dishes Potato dextrose agar medium. The plates were incubated at 28-30°C for 4-7 days. Individual colonies were picked by repetitively streaked through a series of plates to isolate the individual member. For purification, the growing single colonies were picked up and streaked on slants of PDA (plus 4 ml per liter of chloramphenicol, 25 mg/ml, after autoclaving) and kept at 4°C until using.

The isolated colonies were microscopically examined using lactophenol cotton blue technique and identified according to their observed morphological characteristics.

3.3. Identification of Fungal Strains

The isolation of DNA from mycelium was performed according to the method described by (**Ferracin *et al.*, 2009**).Briefly, mycelium from liquid cultures was recovered by filtration and pulverized to a fine powder under liquid nitrogen in a mortar. Approximately 400 mg of the ground mycelium was suspended in 800mL of lysis buffer (200mMTris-HCl; 250mM NaCl; 25mM EDTA; 1% wt/vol sodium dodecyl sulfate) and maintained at 65°C for 20 min. The DNA was purified with phenol:chloroform (25:24) and chloroform:isoamyl alcohol (24:1) and precipitated in 3MNaCl solution in the presence of 95% ethanol, then was hedwith 70% ethanol, and resuspended in ultrapure water. For the identification of *A. niger* and *A. welwitschiae* isolates, two species-specific primers have been designed based on the alignment of partial calmodulin sequences of the members of section Nigri (**Gherbawy *et al.*, 2015**). The first primer was awaspec (ATTTCGACAGCATTTCTCAGAATTA) and the second primer was nigspec (GACAGCATTTTCCAGAACGA). The designed primers were used in combination with the common primer cmd6

(CCGATAGAGGTCATAACGTGG) described by (**Hong *et al.*, 2006**). Amplification was conducted in a thermal cycler with an initial denaturation of 2min at 94°C,followed by 30 cycles of 35 s at 94°C, 20 s at 56°C, 30 s at72°C, and a final extension of 1 min at 72°C. Aliquots of PCR products were checked by electrophoresis on a 1.4% agarose gel revealed with ethidium bromide and visualized by UV transillumination.

3.4. Fungal Cultures Maintenance

Aspergillus niger strains used in the study include the local isolates which were obtained from various locations in Riyadh region in addition to fungal strain, *Aspergillus niger EMCC132* that used throughout the experiments as a reference strain for citric acid production and it was obtained officially from Cairo Microbiological Resources Centre (Cairo MIRCEN), Ain Shams University, Cairo, Egypt.

The cultures of *Aspergillus niger* were maintained on sterilized potato dextrose agar medium (PDA) and stored at 5°C in the refrigerator. In general, all the used media, unless otherwise stated, were sterilized in autoclave at 15-lbs/inch^2 pressure (121°C) for 15 min.

3.5. Screening of Local Fungal Strains for Citric Acid Productivity

The obtained cultures of *Aspergillus niger* were restricted depending on their capability for citric acid production using dye method as described by **Ali (2004) and Kareem *et al.*, (2010)**. The *A. niger* cultures were qualitatively screened in Petri plates containing Czapek-Dox agar medium with Bromocresol green as an indicator (Table 3.1). The medium was prepared by dissolving all of the ingredients except agar; the pH was maintained at 6.0. Then, the medium was sterilized in the autoclave after the agar addition.

Table (3.1): Composition of Czapek-Dox agar medium.

Component	Per liter
Sucrose	30.0 g
Sodium Nitrate ($NaNO_3$)	3.0 g
Dipotassium Phosphate (K_2HPO_4)	1.0 g
Magnesium Sulfate ($MgSO_4.7H_2O$)	0.5 g
Potassium Chloride (KCl)	0.5 g
Ferrous sulfate ($FeSO_4$)	0.01 g
Bromocresol green	5 ml (1.0 % w/v water)
Agar	20.0 g

About 10 ml of the prepared Czapek-Dox agar medium was poured onto sterile Petri Dish and left awhile for solidification at room temperature. Subsequently, 6 mm disc taken from the edge of activity growing colonies of the tested fungal strains was aseptically transferred and placed in the center of medium agar plates.

All inoculated plates of tested were incubated at 28 ± 2 °C for 5 days. Diameter of formatted yellow zones, due to citric acid release, was measured and the values were expressed in millimeters diameter. On the basis of diameter or the widest yellow zones, *Aspergillus niger'* isolates were selected, picked and transferred to the PDA slants for further screening by submerged fermentation in shake flasks.

3.6. Preparation of Conidial Inoculums

The vegetative spore suspensions of fungal strains were prepared by washing 3 days old culture slants with sterilized saline solution (0.85 % NaCl) containing 0.01% Tween 80, gently agitating the tube for 1 min to remove the conidia to obtain conidial suspension. For conidial count or density, the resulting spores' suspension was collected in a sterile tube and the spores were counted using a Neubauer haemocytometer. The conidial count in 1.0 ml of inoculums was calculated after counting the conidia in a square (0.1 mm depth) under microscope. Sufficient saline solution was added to the fungal spores' suspension to give a count within rang approximately $0.6\text{-}1.0 \times 10^6$ spores/ml.

3.7. Citric Acid Production in Submerged Culture (shake-flask culture)

The medium used was Czapek-Dox broth medium which contains (g/l): Sucrose, 30.0; NaNO$_3$, 3.0; K$_2$HPO$_4$, 1.0; MgSO$_4$.7H$_2$O, 0.5; KCl, 0.5; FeSO$_4$, 0.01 and the pH was adjusted to 6.50. The experiments were performed in shake-flask culture using 50 ml of Czapek-Dox broth medium contained in 250 ml Erlenmeyer flask. All media were sterilized at 121°C for 15 min after cooling at room temperature and seeded with 1ml of inoculums (conidia suspensions). The flasks were incubated in a rotary shaker operating 100 rpm at 28°C.

After incubation periods, the culture broth from each flask was filtered through Whatman filter paper for mycelia separation, dried and the yield of dry fungal biomass was recorded. The clear culture filtrate was employed for determination of citric acid and recorded final pH values.

3.8. Time Course for Citric Acid Production

Six periods ranging from 1 to 6 days were tested. After 24 hours from zero time, citric acid produced in cell free extract, final pH and fungal biomass, were determined. These analyses were repeated with different fermentation periods *viz.*, 24, 48, 72, 96, 120, 144 hrs respectively, to determine which period was optimum to produce the highest productivity of citric acid for each tested fungal isolates comparing with reference fungal strain.

3.9. Optimization of Citric Acid Production Parameters

The effect of certain physical factors (agitation or aeration , temperature, pH), biological factor (inoculums density) and nutrients requirements (source and concentration of carbon, source and concentration of nitrogen source, phosphorus source, source and concentration alcohol) on citric acid production were studied by adjusting these parameters or adding these nutrients to the basal medium. Optimization of these various parameters was carried out with "one at a time" strategy keeping all other variable constant except one. Then, the best parameter was used for the further experiment by which the acid production improved.

3.9.1. Agitation (Static or Shaking Culture)

To test the effect of shaking on the production of citric acid by selected fungal isolates and the reference fungal strain, two sets of flasks with Czapek-Dox broth medium were prepared and inoculated with the prepared inocula. One set was incubated without shaking and the second set kept continuously on a rotary shaker (100 rpm). Both groups were incubated at 28 °C for 5 days.

3.9.2. Incubation Temperature

To investigate the optimum incubation temperature, assays were carried out at different temperatures (12, 20, 28, 36 and 44°C). Flasks contain 50 ml of Czapek-Dox broth medium were inoculated with isolated fungal, 1 ml of spores suspension used as inoculums. The flasks were incubated in rotary shaker operating at 100 rpm at the different temperatures for 5 days. After that the mycelia separation by filtration to assay the biomass dry weight and the filtrate was analyzed for pH and citric acid amounts.

3.9.3. Initial pH Value

To determine the impact of initial pH on citric acid production, 50 ml of Czapek-Dox broth medium in 250-ml Erlenmeyer flasks was prepared and adjusted at various pH values (3, 5, 8 and 10) and the control one was pH 6.5, to find out the optimum pH value for production of citric acid. These pH values were adjusted by using concentrated sodium hydroxide and hydrochloric acid. The sterilized flasks were inoculated with 1 ml of spore's suspensions as inoculums of tested fungal isolates, and then incubated in rotary shaker operating at 100 rpm at 28 °C for five days.

3.9.4. Inoculum Size

Experiments were performed in shake-flask culture using 50 ml of Czapek-Dox broth medium in 250-ml Erlenmeyer flasks and pH was adjusted at optimum value. The sterilized flasks were inoculated with 1 ml of spore's suspensions as inoculums of tested fungal isolates. Different spore suspensions (1.0, 2.0 and 3.0) $x10^6$ spores/ml were used and the flasks were incubated in rotary shaker operating at 100 rpm at 28 °C for five days.

3.9.5. Carbon Source

To determine the best carbon source for the production of citric acid, experiments were performed in shake-flask culture using 250-ml Erlenmeyer flask contained in 50 ml of Czapek-Dox broth medium. Where the tested fungal strains were grown in the broth medium containing different carbon sources, including sucrose, maltose, glucose, fructose, lactose and arabinose which were individually added in the same amount of the initial concentration of carbon source (3% w/v) in broth medium. Before sterilization, pH values were adjusted at the optima pH value. The sterilized flasks were inoculated with fungal inuculom, 1 ml of optimum density of the spores suspension used as inoculums. The flasks were incubated in rotary shaker operating at 100 rpm at 28 °C for 5days.

3.9.6. Concentration of Carbon Source

Experiments were performed in shake-flask culture using 250-ml Erlenmeyer flask contained 50 ml of Czapek-Dox broth medium amended with different concentrations of favorite carbon source (20, 25, 30, 35, 40 g/l). The sterilized flasks were inoculated with 1 ml of spore's suspensions as inoculums of tested fungal isolates and incubated in rotary shaker operating at 100 rpm at 28 °C for five days.

3.9.7. Nitrogen Source

To determine the best nitrogen source for citric acid production, the selected fungal strains was grown in shake-flask culture using 250-ml Erlenmeyer flask contained in 50 ml of modified Czapek-Dox broth medium amended with various nitrogenous supplements, the organic sources were peptone, urea, beef extract and the inorganic source were ammonium nitrate NH_4NO_3, ammonium sulfate, $(NH_4)_2SO_4$ and ammonium chloride (NH_4Cl) which were added to the medium by replaced sodium nitrate $NaNO_3$ and added as 3.0 g/l from each nitrogen source. The flasks were inoculated with isolated fungal, 1 ml of spores suspension optimum density used as inoculums. The flasks were incubated in rotary shaker operating at 100 rpm at 28°C for 5days.

3.9.8. Concentration of Nitrogen Source

Experiments were performed in shake-flask culture using 250-ml Erlenmeyer flask contained 50 ml of modified Czapek-Dox broth medium amended with different concentrations of favorite nitrogen source (1, 2, 3, 4, 5 g/l). The sterilized flasks were inoculated with 1 ml of spore's suspensions as inoculums of tested fungal isolates and incubated in rotary shaker operating at 100 rpm at 28 °C for five days.

3.9.9. Phosphorus Source

To determine the favorite phosphorus source for the production of citric acid, experiments were performed in shake-flask culture using 250-ml Erlenmeyer flask contained in 50 ml of modified Czapek-Dox broth medium. Where the tested fungal strains were grown in the broth medium containing different phosphorus sources, including NaH_2PO_4, Na_2HPO_4, Na_3PO_4, KH_2PO_4 which were individually added in the same amount of the initial concentration of phosphorus source, K_2HPO_4 (1.0 g/l) in broth medium. Before sterilization, pH values were adjusted at the optima pH value. The sterilized flasks were inoculated with fungal inuculom, 1 ml of optimum density of the spores suspension used as inoculums. The flasks were incubated in rotary shaker operating at 100 rpm at 28 °C for 5days.

3.9.10. Enhancement by Alcohol

To assess the effect of alcohol on citric acid production, 1.0 percent of different alcohol sources (Methanol, Propanol and Ethanol) were added to 250-ml Erlenmeyer flask contained in 50 ml of modified Czapek-Dox broth medium after autoclaving. Then, flasks were inoculated with fungal inoculums, 1 ml of optimum density of the spores suspension used as inoculums. The flasks and their content were then incubated in rotary shaker operating at 100 rpm at 28 °C for 5days.

3.9.11. Concentration of Alcohol

Experiments were performed in shake-flask culture using 250-ml Erlenmeyer flask contained 50 ml of modified Czapek-Dox broth medium amended with different concentrations of the best alcohol source (1.0, 2.0, 3.0, 4.0, 5.0%) after sterilization. Then, the flasks were inoculated with 1 ml of spore's suspensions as inoculums of tested fungal isolates and incubated in rotary shaker operating at 100 rpm at 28 °C for five days.

3.12. Usage of Raw Materiales for Citric Acid Production

3.12.1. Raw Material Sources

Five different agricultural waste products (pineapple, sugar cane bagasse, sugar cane molasses, dates molasses, potato peels) used in this study was kindly obtained from various sources, e.i., sugarcane bagasse from sugarcane juice shops; potato peel from Saudi factory for the manufacture of potato and food – Riyadh - third industrial region; pineapple peel from vegetables markets – alraboa – Riyadh; dates molasses from Tuwaijri farm – Qassim and sugar cane molasses from the United Company for sugar – Jeddah.

3.12.2. Preparation of Raw Materials

Samples of sugar cane bagasse, potatoes peels and pineapple peels were dried in outside air (at the roof) for a week with a daily continuous stirring. After the samples completely dried, were milled using blinder, then sieved using sieve with narrowness holes. All samples were kept in clean bottles at a temperature of 25-28 until use

3.12.3. Chemical Analysis of the Raw Materials

All wastes samples were dried in oven at 105°C. For carbon, hydrogen, nitrogen and sulfur (CHNS) analysis, the samples were passed through 75 Micron Sieve and CHNS were analyzed by high combustion method using PERKIN ELMER Series II, CHNS/O Analyzer. Phosphorus (soluble) was analyzed by Ascorbic Acid method using HACH DR/U 4000 Spectrophotometer. Soluble components of Potassium, Calcium and Magnesium were analyzed by Ion Chromatography using Shimadzu Ion Chromatograph.

3.12.4. Raw Materials and Saccharification

Referring to the chemical analysis of used raw materials (Table 7.1 in Appendix) the saccharification of waste materials used for the production of citric acid consisted of 5% in distilled water (w/v or v/v) of selected raw materials (i.e. sugarcane bagasse, pineapple peel, dates molasses and sugarcane molasses) except only 3% (w/v) of potato peel because of high viscosity of the suspended materials. Autoclaving (sterilization) of all prepared solutions were conducted for mild hydrolysis the raw materials. After cooling, the suspended insoluble materials were removed by filtration and the obtained filtrates were used in preparation of fermentation media for citric acid production.

3.12.5. Raw Materials and Production Medium

Fermentation experiments were performed in 250 ml Erlenmeyer flasks containing Czapek-Dox broth medium components dissolved in 50 ml of the previous prepared filtrate (5 or 3% of the waste), then they were cotton plugged and sterilized at 15 lbs for 15 minutes at 121° C. Flasks were inoculated with isolated fungal, 1 ml of spores suspension used as inoculums. The flasks were incubated in rotary shaker operating at 100 rpm at the different temperatures for 5 days. After that the mycelia separation by filtration to assay the biomass dry weight and the filtrate was analyzed for pH and citric acid amounts.

3.13. Mutagenesis of Fungal Strains

The three selected isolates of *Aspergillus nigar* were grown for 4-5 days at 30°C on Potato Dextrose Agar (PDA) slants in triplicate.

Completely grown slants were maintained in the refrigerator until further use. A CAMAG UV chamber was used for mutation (physical, UV and chemical, NTG).

Preparation of sample: The inoculated broths were centrifuged at 3000 g for 10 min under cooling and the supernatant was diluted 1:50 for citric acid determination. The dilutions were membrane filtered (0.45 μm) before injection.

Chemicals: Citric acid was purchased from Merck and Sigma-Aldrich. Acetonitrile, potassium dihydrogen orthophosphate and phosphoric acid were of analytical purity or for chromatographic use. The water used was ultrapure water.

Organic acid standards: A standard stock solution was prepared containing 1000 mg/l citric acid. The stock solution and the corresponding dilutions was made in ultrapure water and stored in dark places between the experiments, at low temperature (+4°C).

3.13.1. Physical Mutation

Young spores, 16-18 h old, were harvested using 0.1% Tween-80. Then 1 ml samples of the spore suspension containing approximately 1×10^6

cells ml^{-1} were subjected to UV irradiation for different time exposures(60, 120 and 180 min) in a sterile Petri dish using a CAMAG UV chamber, which emits radiations of 254 and 365 nm at a distance of 25 cm according to **Usha *et al.,* (2001)**. The experiment was carried out in triplicate. The spore suspensions were intermittently agitated during the course of irradiation. The treated spore suspensions were then incubated at 4°C in the dark overnight for any DNA repair to take place. Suspensions were suitably diluted and appropriate dilutions were plated on PD broth and incubated for 4-5 days at 30°C. The suspensions were centrifuged under cooling to estimate biomass and calculated the percentage of citric acid by HPLC protocol.

3.13.2. Chemical Mutation

The pure substance of N-methyl-N-nitro-N-nitrosoguanidine (Sigma) was dissolved in different volumes of 0.05M of citric buffer (pH 5) to form 100, 200, 400, 600 and 750 µg ml^{-1} concentration.

Spore suspension:

Young spores of tested isolates of *A. niger*, 16-18 hours old, were harvested using distilled and sterilized water. Then 1 ml samples of the spore suspension containing approximately 1×10^{7} spores ml^{-1} were subjected to different concentration of NTG for different time exposures (30 and 60 min) in a sterile flasks which kept in water bath at 24°C according to **Mutwakil (2011)**.

After interval time, the samples were centrifuged to collect cells and many serial dilutions were done. The treated spore suspensions were plated on PD broth and incubated for 4-5 days at 30°C. The PD broth cultures were centrifuged under cooling to estimate fungal biomass and determine the citric acid concentration in supernatant by HPLC.

3.14. Assay Methods

3.14.1. Citric Acid Estimation

Citric acid levels were measured by two techniques, colorimetric and HPLC (high performance liquid chromatography) methods.

3.14.1.1. Colorimetric Method

Citric acid levels were determined calorimetrically following the recommended pyridine-acetic anhydride method (**Marier and Boulet 1958**).

3.14.1.(A) Standard Curve of Citric Acid

Appropriate dilutions (2-20 mg/ml) were made from citric acid anhydrous stock solution (40 mg/ml). One milliliter of each dilution followed by 1.30 ml of pyridine was added into individual test tubes and swirled. Acetic anhydride (5.70 ml) was then added into each tube. All dilution tubes were placed in a water bath at 32°C for 30 min. A blank was run in parallel re   placing 1.0 ml of the sample with distilled water. The optical density was measured at 420 nm on spectrophotometer. A calibration curve was drawn taking the citric acid concentration at X-axis and optical density at Y-axis (Figure 3.1).

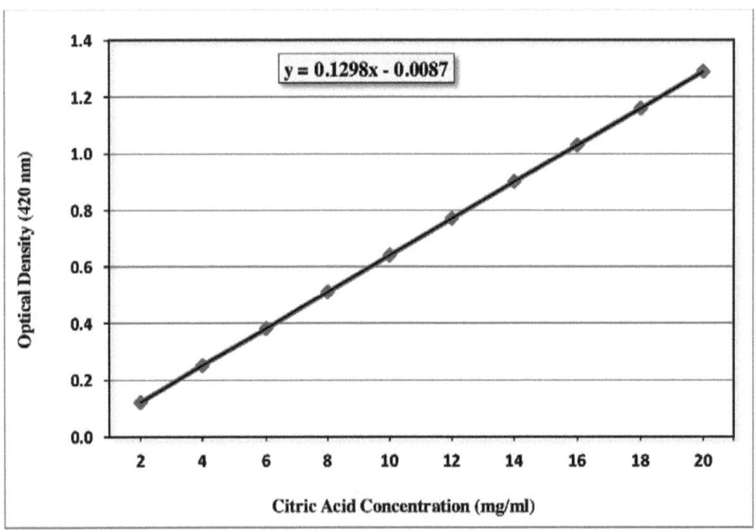

Fig. (3.1): Standard curve of citric acid.

3.14.1.1.(B) .Estimation of Citric Acid in the Culture Filtrate

Citric acid was spectrophotometrically measured according to procedure of **Marier and Boulet (1958)**. One milliliter of culture filtrates (or diluted) along with 1.30 ml of pyridine was added into a test tube and swirled briskly prior to add 5.70 ml Acetic anhydride. Then, the test tube was placed in a water bath at $30 \pm 2°C$ for 30 min. The optical density was measured at 420 nm on spectrophotometer. The citric acid concentration of the sample was estimated from a blank or reference tube (run parallel, replacing 1.0 ml of the culture filtrate with distilled water).

3.14.1. 2. Analysis by HPLC

The citric acids in the sample test solution were separated by reversed phase chromatography on a 250 mm×4.6 mm i.d., 5 µm particle Hypersil Column, of which were detected by absorbance and quantified with external calibration graphs. The detector was set at 214 nm.

The HPLC analysis was performed with a Surveyor Thermo Electron system comprising a vacuum degasser, Surveyor Plus LCPMPP pump, Surveyor Plus ASP autosampler and diode array detector with 5 cm flow cell. Integration, data storage and processing were performed by Chrom Quest 4.2 software according to **Nour *et al.,* (2010)**.

The determinations were made in isocratic conditions, at 10°C, using a mobile phase made of 50 mM phosphate solution (dissolve 6.8 g potassium dihydrogen phosphate in 900 ml water; the pH value should be adjusted to pH =2.8 with phosphoric acid and then filled to 1000 ml with water) filtered through a polyamide membrane (0.2 μm) and degassed in a vacuum. The flow rate of the mobile phase was 0.7 ml/min for all the chromatographic separations. The separation column was balanced with mobile phase until the baseline was stabilized. Sample injections were made at this point. The volume injected was 5 μl for either prepared sample or standard solution.

3.14.2. Mycelial Dry Weight Measurement

Fungal biomass was determined by transferring the mycelium to dried and weighted. The mycelial pellets of each flask were harvested by filtration of the culture medium through pre-weighted filter paper (Whatman paper No.1) under suction. The biomass was determined after washing the mycelial mass with distilled water and dried at 105°C in an electric oven overnight until a constant weight. The biomass dry weight was estimated from the difference between the mycelium plus filter paper and the filter paper itself. The biomass dry weight was expressed in term of mg of the mycelium per 50 ml of the culture volume.

3.14.3. Measurement of pH

The initial pH values and the changes of pH degree in the growing cultures were regularly measured by using electrical pH meter.

3.15. Statistical analysis

Statistical analyses of the data were undertaken using STATISTIX 10 computer program (analytical software). The experiments were analyzed by one-way analysis of variance (AVONA) and the significance of the differences between means was calculating using the Duncan test. Means values were separated on the basis of least significant difference (LSD) at the 0.05 probability level. All values given in the present research are means of three replicates. The results are presented as mean ± standard error (SE).

4. RESULTS AND DISCUSSION

4.1. First Part: Isolation of *Aspergillus niger* and preliminary screening for their capability of citric acid production

Citric acid is one of the most common products which have a never ending demand in the global market. Citric acid fermentation is one of the primitive fermentations but still its production is increasing with passage of time. In 2007, its global production has exceeded 1.6 million tons (**Pandey** ***et al.*, 2013**).

Aspergillus niger is an important industrial organism, and has been the subject of research for several decades. Since *A. niger* holds the GRAS (Generally Regarded As Safe) status with the US Food and Drug Administration (FDA), it is the preferred production organism for a range of different products. Since the 1960's, it has been exploited in the food industry, especially for its ability to produce a vast range of acids, all industrially important and used in many applications in today's modern life (**Abarca *et al.*, 2004**).

Nowadays the most economical technology for the production of citric acid uses carbohydrates fermented by bacteria, fungi and yeast, under submerged conditions. A large number of fungal species have the ability to produce citric acid. Although many microorganisms can be used to produce citric acid, *Aspergillus niger* remains the main organism of choice for commercial production and industrial producer of citric acid for many reasons. The main advantages of using *A. niger* are: its production more citric acid per time unite, its high yield, its ease of handling and its ability to ferment a variety of cheap raw materials (**Socco *et al.,* 2006**).

4.2.1. *Aspergillus niger* isolation

The aim of this part is to isolate and get pure cultures of local fungus, *A. niger'* strains. From twenty soil samples collected from different sites in Riyadh region (as shown in Appendix), several *Aspergilli* strains were isolated and purified.

Among the colonies characteristics observed, the identification process was performed by lacto phenol cotton blue method based on their morphological characteristics. *Aspergillus niger* which shows black with white margin (as shown in Appendix), colorless brownish shade conidiophores, globosely, thick walled brown color conidia (as shown in Appendix), it was confirmed by Gillman fungi Manual.

Based on morphological characterization, ninety six isolates of *Aspergillus niger* from different sources were selected as mentioned in Table (4.I.1), the twenty sites and number of *A. niger* isolates from each site with an abbreviation cods.

Table (4.I.1.): Sources of ninety six *Aspergillus niger'* isolates

No.	Soil Source	Isolates Codes (No.)
1.	South Wadi Hanifa	SWH (6)
2.	North Wadi Hanifa	NWH (5)
3.	Slam Park in the center of Riyadh	SPR (7)
4.	King Saud University Malaz‘ Garden	KSUM (5)
5.	King Saud University Diriya‘ Garden	KSUD (5)
6.	Rawabi garden east of Riyadh	EGR (4)
7.	Farm in Thumama (A)	FTHA (5)
8.	Farm in Thumama (B)	FTHB (3)
9.	Farm way Qassim (A)	FQWA (5)
10.	Farm way Qassim (B)	FQWB (3)
11.	Farm in Mansuriya	FMN (5)
12.	A Farm in AL-Muzahmiyya	FMZ (6)
13.	A Farm in AL-Ghad	FGD (5)
14.	A Farm in Dirab (A)	FDRA (6)
15.	A Farm in Dirab (B)	FDRB (6)
16.	A Farm in the dam Diriya	FDD (4)
17.	A Farm in AL-Kharj	FKH (4)
18.	A Farm in Dharme	FDR (4)
19.	Farm in Hayer	FHA (4)
20.	Farm in Arka	FAR (4)

4.2.2. Preliminary screening of *A. niger* isolates producing citric acid

The experiment was conducted to choose citric acid producing-fungal isolates using plates of Czapek-Dox agar containing Bromocresol green as an indicator and yellow zones indicated citric acid production.

To check from efficacy of the obtained isolates for citric acid production, the experiment was conducted using plates of Czapek-Dox agar containing Bromocresol green as an indicator and yellow zones indicated citric acid production.

Ninety six isolates of *Aspergillus niger* were tested for their capability to produce citric acid using an indicator plates medium. Detailed data recorded in Table (4.I.2) obviously indicated that all *A. niger* isolates understudying were varied in their capability to produce citric acid and in productivity amount at different time intervals i.e. 24, 48, 72, 96 hours incubation on indicator plats medium through the diameter of the formed yellow zone (mm).

Data recorded in Table (4.I.2), diagramed in Figure (4.I.3) and photographed in Figure (4.I.4) indicate that most of *A. niger* isolates (70 isolates) reached to the maximum acid production, where yellow color filled the plates medium (90 mm), after 72 hour incubation, represented 72.9 % of the total isolates (Fig. 4.I.3). However, some isolates (14 isolates) were slowly released acid in the indicator medium and acid production gave a maximum yellow zone diameter (90 mm) after 96 h incubation, represented 14.6 % of the total isolates. In the same time, twelve *A. niger* isolates proved to be the best citric acid producters, where gave a widest yellow zone diameter (90 mm) after only 48 hour incubation, represented 12.5 % of the total isolates (Fig. 4.I.3).

On the basis of obtained results of the preliminary screening through used an easily and quick technique, It could be able to select six *A. niger* isolates, that obtained from different isolation sources, gave highest productivity of citric acid, ranged from 49 to 60 mm zone diameter, after 24 hours incubation.

The favorable isolates of *A. niger* for citric acid production were SWH4, KSUD4, FTH1, FQWA1, FDRA3 and FAR4 selected for further experiments.

In view of other workers, the citric acid is being produced by microorganisms during fermentation through their metabolic citric acid cycle (**Haq *et al.*, 2002 and Ali 2004**). When microbes concern, there are many organisms which are producing citric acids. However, fungus is considered to be suitable producer of citric acid, especially, *Aspergillus niger* is utilized as much as it produces.

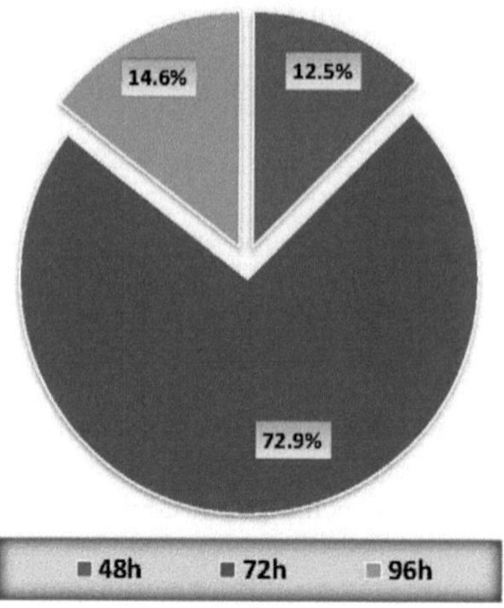

Fig. (4.I.3): Distribution of all *A. niger* isolates (96) producing citric acid after different incubation periods (hours).

Table (4.I.2): Screening of *Aspergillus niger* isolates for citric acid Production after various incubation periods using indicator plates medium (as yellow zone diameter, mm).

No.	Region Code	No.	Fungal Isolates	Yellow Zone Diameter (mm)			
				24 h	48 h	72 h	96 h
1.	SWH	1.	SWH1	49	90		
		2.	SWH2	39	75	90	
		3.	SWH3	49	68	90	
		4.	SWH4	57	90		
		5.	SWH5	27	56	90	
		6.	SWH6	19	39	65	90
2.	NWH	7.	NWH1	16	49	64	90
		8.	NWH2	13	52	90	
		9.	NWH3	24	64	90	
		10.	NWH4	39	58	90	
		11.	NWH5	25	57	90	
3.	SPR	12.	SPR1	45	77	90	
		13.	SPR2	46	72	90	
		14.	SPR3	47	90		
		15.	SPR4	32	54	90	
		16.	SPR5	30	55	90	
		17.	SPR6	53	90		
		18.	SPR7	24	56	90	
4.	KSUM	19.	KSUM1	14	50	90	
		20.	KSUM2	15	48	90	
		21.	KSUM3	17	43	90	
		22.	KSUM4	42	62	90	
		23.	KSUM5	16	49	90	
5.	KSUD	24.	KSUD1	43	63	90	

No.	Region Code	No.	Fungal Isolates	Yellow Zone Diameter (mm)			
				24 h	48 h	72 h	96 h
		25.	KSUD2	57	79	90	
		26.	KSUD3	37	63	90	
		27.	KSUD4	60	90		
		28.	KSUD5	34	59	90	
6.	EGR	29.	EGR1	29	55	90	
		30.	EGR2	25	45	65	90
		31.	EGR3	17	47	90	
		32.	EGR4	16	52	90	
7.	FTHA	33.	FTHA1	55	90	90	
		34.	FTHA2	52	78	90	
		35.	FTHA3	35	70	90	
		36.	FTHA4	48	65	90	
		37.	FTHA5	zero	35	52	90
8.	FTHB	38.	FTHB1	39	65	90	
		39.	FTHB2	47	78	90	
		40.	FTHB3	40	78	90	
9.	FQWA	41.	FQWA1	55	90		
		42.	FQWA2	39	72	90	
		43.	FQWA3	48	70	90	
		44.	FQWA4	55	90		
		45.	FQWA5	43	63	90	
10.	FQWB	46.	FQWB1	25	59	90	
		47.	FQWB2	15	49	90	
		48.	FQWB3	zero	30	5.7	90
11.	FMN	49.	FMN1	37	70	90	
		50.	FMN2	14	58	90	
		51.	FMN3	15	60	90	
		52.	FMN4	15	67	90	

No.	Region Code	No.	Fungal Isolates	Yellow Zone Diameter (mm)			
				24 h	48 h	72 h	96 h
		53.	FMN5	Zero	47	90	
12.	FMZ	54.	FMZ1	Zero	42	68	90
		55.	FMZ2	15	4.9	70	90
		56.	FMZ3	23	54	90	
		57.	FMZ4	24	48	90	
		58.	FMZ5	Zero	37	72	90
		59.	FMZ6	15	52	68	90
13.	FGD	60.	FGD1	51	7	90	
		61.	FGD2	35	67	90	
		62.	FGD3	36	53	90	
		63.	FGD4	27	55	90	
		64.	FGD5	53	74	90	
14.	FDRA	65.	FDRA1	30	58	90	
		66.	FDRA2	27	74	90	
		67.	FDRA3	59	90		
		68.	FDRA4	15	53	75	90
		69.	FDRA5	30	65	90	
		70.	FDRA6	59	90		
15.	FDRB	71.	FDRB1	13	57	90	
		72.	FDRB2	Zero	43	73	90
		73.	FDRB3	52	90		
		74.	FDRB4	15	60	90	
		75.	FDRB5	Zero	64	90	
		76.	FDRB6	Zero	44	7	90
16.	FDD	77.	FDD1	18	45	6.8	90
		78.	FDD2	34	70	90	
		79.	FDD3	15	47	74	90
		80.	FDD4	20	59	90	

No.	Region Code	No.	Fungal Isolates	Yellow Zone Diameter (mm)			
				24 h	48 h	72 h	96 h
17.	FKH	81.	FKH1	38	75	90	
		82.	FKH2	31	60	90	
		83.	FKH3	33	73	90	
		84.	FKH4	48	73	90	
18.	FDR	85.	FDR1	33	74	90	
		86.	FDR2	14	60	90	
		87.	FDR3	10	50	90	
		88.	FDR4	38	71	90	
19.	FHA	89.	FHA1	39	64	90	
		90.	FHA2	20	64	90	
		91.	FHA3	15	68	90	
		92.	FHA4	15	63	90	
20.	FAR	93.	FAR1	18	69	90	
		94.	FAR2	18	58	90	
		95.	FAR3	38	72	90	
		96.	FAR4	55	90		

Highest Category (12 isolates reached max. acid release after 48 h)

Med. Category (70 isolates reached max. acid release after 72 h)

Lowest Category (14 isolates reached max. acid release after 96 h)

Fig. 4.I.4. Detection for citric acid production on indicator plate's medium for various *A. niger* isolates at different time intervals.

4.2.3. Identification of citric acid producing *Aspergillus niger'* isolates

Minty six fungal isolates were identified conventionally referencing to their macroscopic characteristics (colonial morphology, colour and appearance of colony, shape and diameter of colony) and microscopic characterization (septation of mycelium, shape, diameter and texture of conidia).

All the isolates identified as filamentous fungi belonging to the phyla *Deuteromycota*. According to the features represented in photographs images Appendix (Fig. 7.4 & 7.5), they were identified to the genus level as black *Aspergillus*.

From the nine six black *Aspergillii* only two isolates (FQW and KSUD comparing to the reference strain *A. niger* EMCC132) were identified at the species level depending on both morphological and molecular characteristics presented in Figs. (4.I.5 - 4.I.8).

From the obtained data, it was found that KSUD was identified as *A. welwitschiae* in addition to both FQW and EMCC132 proved to be *Aspergillus niger* species complex.

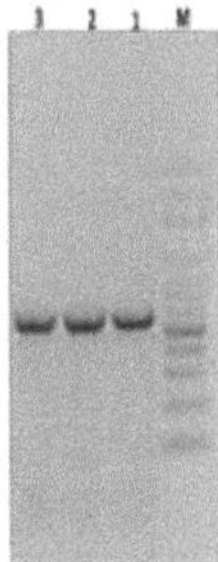

Fig. 4. 1.5 Polymerase chain reaction products obtained amplifying *Aspergillus niger* ITS regions of strains KSUD N, (lane 1), FQW N (lane 2) and EMCC132 N (lane 3) genomic DNA with ITS1 and ITS4 primers. Lane M is a 100 bp DNA ladder.

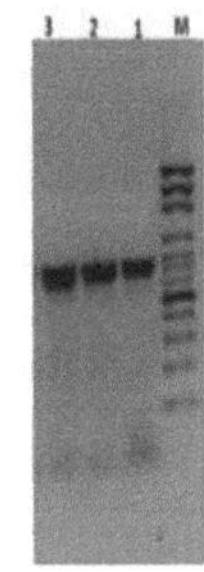

Fig. 4. 1.6. Polymerase chain reaction products obtained amplifying *Aspergillus niger* Beta- tubulin gene of strains KSUD N, (lane 1), FQW N (lane 2) and EMCC132 N (lane 3) genomic DNA with BT2A and BT2B primers. Lane M is a 100 bp DNA ladder.

Fig. 4.1.7. Polymerase chain reaction products obtained amplifying *Aspergillus niger* calmodulin gene of strains KSUD N, (lane 1), FQW N (lane 2) and EMCC132 N (lane 3) genomic DNA with CF2F and CF4R primers. Lane M is a 100 bp DNA ladder.

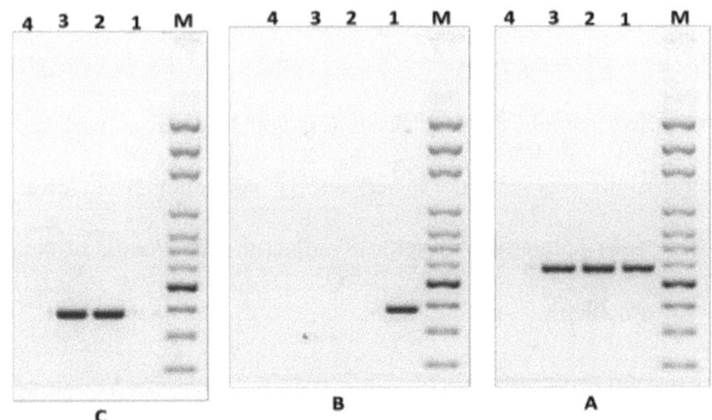

Fig.4.I.8 Polymerase chain reaction products obtained amplifying *Aspergillus niger* strains KSUD N, (lane 1), FQW N (lane 2) , EMCC132 N (lane 3) and *Aspergillus flavus* (lane 4) genomic DNA with Cmd6 and primer 1 (A), Cmd6 and primer 2 (B) and Cmd6 and primer 3 (C). Lane M is a 100 bp DNA ladder.

4. 2. Second Part: Selection and Time Course

One of the most important fungi used in industrial microbiology, *Aspergillus niger*, has been employed for many years for the commercial production of citric acid. However, the worldwide demand for citric acid is increasing faster than its production and more economical processes are required (**Vandenberghe *et al.*, 1999**).

Citric acid production synthesis by fermentation is the most economical and widely used way of obtaining this product. More than 90 % of the citric acid produced in the world is obtained by fermentation, which has its own advantages: operations are simple and stable, the plant is generally less complicated and needs less sophisticated control systems, technical skills required are lower, energy consumption is lower and frequent power failures do not critically affect the functioning of the plant (**Socco *et al.*, 2006**).

Especially, the submerged technique is widely used for citric acid production. It is estimated that about 80 % of world production is obtained by submerged fermentation. This fermentation process employed in large scale requires more sophisticated installations and rigorous control. On the other hand, it presents several advantages such as higher productivity, lower labour costs and lower contamination. Submerged fermentation can be carried out in batch, fed batch or continuous systems, although the batch mode is more frequently used. Normally, citric fermentation is concluded in 5 to 12 days, depending on the process conditions (**Ambati and Ayyanna 2001**).

In addition, the submerged process has become the method of choice in the industrialized countries, because it is less labour intensive, yields higher production rate, and uses less space (**Prasad** *et al.,* **2014**).

As well, the choice of a suitable strain is of great importance. Among the different fungal isolates of *A. niger* has the highest ability to produce citric acid (**Ujcova** *et al.,* **1980**).

4. 2.1. Selection of Citric Acid Producers

Therefore in this part, the best six citric acid producing-fungal isolates (which selected after primary qualitatively screening for their potential as citric acid producers on plates of Czapek-Dox agar with Bromocresol green as an indicator) were selected from a group of indigenous *A. niger* population to study their capability for citric acid production in submerged culture, shake flask cultures, during three days incubation.

The selected six *A. niger* isolates are FDR, SWH, FAR, FQW, KSUD and FTH. In addition to *A. niger* EMCC132 was used as a reference fungal strain for citric acid production.

Data recorded in Table (4.II.1) and graphed in Fig. (4.II.1) shows that the production of citric acid by six locally isolated and selected reference strain *A. niger* EMCC132 were significantly varied. Citric acid production was in the range from 4.52 to 16.34 mg/10 ml culture filtrate. A highest significant citric acid concentration (16.34 mg/10ml) was produced by *A. niger* EMCC132 followed by KSUD, FQW and FDR that produced not significantly different amounts of citric acid, 9.98, 9.07 and 8.84 mg/10ml, respectively.

Dry fungal mass measurements were ranged from 0.24 to 0.41 g/50 ml. It noticed that no correlation between fungal biomass of all tested isolates and produced citric acid amounts (Table 4.II.1).

The pH measurements of the tested fungal cultures after 3 days incubation reduced and reached to acid range (3.19 − 3.63) as shown in Table (4.II.1).

The submerged fermentation process is desirable and fermentative production of citric acid arises from a primary energy metabolism although it is non-growth associated (**Rane and Sims 1996**).

Based on the productivity of citric acid under these conditions, higher two local *A. niger* isolates, KSUD and FQW, were selected for further experiments comparing to the reference strain, *A. niger* EMCC132.

Table (4.II.1): Production of citric acid (mg/10ml) by most potent isolates of *Aspergillus niger* comparing with reference stain (EMCC132) in submerged culture. Data represented as mean ± standard error.

Isolate	pH	Growth (DW g/50 ml)	Citric Acid (mg/10ml)
EMCC132	$3.61^a \pm 0.06$	$0.29^{ac} \pm 0.03$	$16.34^a \pm 0.09$
FDR	$3.26^{cd} \pm 0.06$	$0.41^a \pm 0.05$	$8.84^b \pm 0.04$
SWH	$3.34^{bc} \pm 0.03$	$0.38^{ab} \pm 0.03$	$6.76^c \pm 0.10$
FAR	$3.49^{ab} \pm 0.12$	$0.28^{bc} \pm 0.01$	$7.05^c \pm 0.33$
FQW	$3.63^a \pm 0.02$	$0.40^{ab} \pm 0.06$	$9.07^b \pm 0.29$
KSUD	$3.19^c \pm 0.04$	$0.24^c \pm 0.06$	$9.98^b \pm 0.13$
FTH	$3.37^{bd} \pm 0.02$	$0.38^{ab} \pm 0.01$	$4.52^d \pm 0.09$
LSD $_{0.05}$	0.1789	0.1207	1.2674

In Czapek-Dox broth medium. Initial pH was 6.5. Inoculum density was ~1.0×10^6 spore/ml. Cultures were grown with shaking at 100 rpm at 28°C for 3 days.
Means in the same column followed by the same letter are not significantly different based on LSD at p = 0.05 according to Duncan's multiple range test.

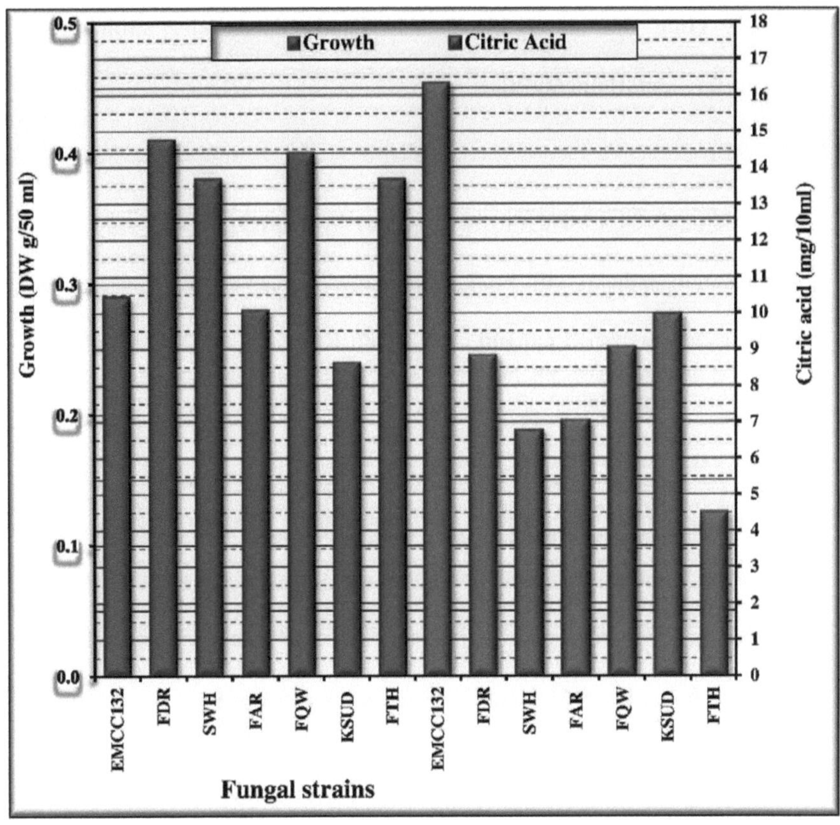

Fig. (4.II.1): Production of citric acid by the most potent strains of *A. niger* comparing with reference stain (EMCC132) in shake submerged culture after 3 days incubation.

4. 2.2. Time Course of Citric Acid Production under submerged fermentation condition

Although citric acid can be produced by submerged and solid state fermentation, industry still prefers submerged fermentation because of its continuous operation (**Leangon *et al.,* 2000; Kumar *et al.,* 2003**). For that, submerged fermentation in shaking flask cultures was used in all over current experiments for citric acid production.

To determine the effect of fermentation period on citric acid productivity and to define the optimum incubation time, fermentation was carried out for various time periods (1 – 6 days). Shaking flask cultures were incubated at 28°C. Samples were withdrawn every 24 hours. Citric acid production, change in pH and dry fungal biomass were also determined.

The obtained results of citric acid production, pH values and fungal biomass that determined in shake flask cultures of the three molds under investigation (*A. niger* KSUD, FQW and EMCC132) are expressed in Table (4.II.2) and graphically illustrated in Figure (4.II.2).

The obtained data show that citric acid amounts produced by all *A. niger* tested isolates were gradually and significantly increased by increment of the incubation periods.

A. niger FQW was grown on Czapek-Dox broth medium and gave maximum biomass, 0.397 mg DW/50 ml, at the 6th day of incubation. In the same time, citric acid was detected and quantified starting from the 1st day and reached to a statistically maximum value, 13.19 mg/10ml, at 5th day shaking incubation (Table 4.II.2 and Fig. 4.II.2). Also the fungal biomass was grown progressively and recoded highest growth after 6 day incubation that was 0.40 g DW/50 ml. In contrast, pH value was gradually reduced that recorded lowest value, 2.29, at the 6th day.

The recorded results in Table (4.II.2) and graphic in Figure (4.II.2) indicate that the time course analysis of submerged culture of *A. niger* KSUD was slowly propagated in the first day, then gradually increased and gave a maximum fungal biomass (0.40 g DW/50 ml) in the 4th day of incubation.

Although the fungal growth started reduces at 5th day incubation, citric acid amount increased and assayed statistically high value, 16.91 mg/10ml among 5 days incubation. The estimated pH values appeared to obviously decrease after first 24 hours and gradually reduced in parallel with acid production, that recorded lowest value, 2.43, at the 6th day incubation (as mentioned in Table 4.II.2).

Comparing the previous data of local fungal isolates with the reference one, it was found that *A. niger* EMCC132 was produced highest amount of citric acid and reached to 25.51 mg/10ml at the 6th day incubation (Table 4.II.2 and Fig. 4.II.2), moreover, gave a maximum growth, 0.38 mg DW/50 ml, where pH values gradually reduced and recorded 2.73 after 6 days.

Although, an international strain *A. niger EMCC132*, used as a reference fungal strain for citric acid production, had a pronounced capability to produce highest amount of citric in flask culture after 6 days incubation. However in case of *A. niger* KSUD and FQW, the maximum amounts of citric acid were at the 5th day and the higher amounts of citric acid at sixth day were insignificantly increased. But the variation was not significantly.

Moreover, the quantity of citric acid produced varies with both microorganism and fermentation period. According to the pervious results, it is indicated that among all the tested fungal strains, citric acid production was varied. Citric acid was ranged between 14.09 – 25.51 mg/10ml.

The tested fungal strains were produced citric acid in the following order, *A. niger* EMCC132 > *A. niger* KSUD > *A. niger* FQW.

On comparing the time course of the citric acid production in the fermentation medium to that of incubation period where citric acid amounts reached statistically to maximum values on 5th day of fermentation.

The optimal time of incubation for maximum citric acid production varies with both the organism.

The obtained results of the time course analysis of tested fungal cultures indicated that optimal time incubation for the maximum citric acid production varies with both the tested fungal strains under the same condition and ranged between 5-6 days incubation period.

Similarly, it was found by **Vergano *et al.,* (1996)** and **Vasanthabharathi *et al.,* (2013)** who reported that viability increases with time of incubation, but higher production of citric acid was achieved and reached maximum at the stationery phase in less than 7 days incubation.

While study of **Maharani et al., (2014)** revealed that the optimum time course is 192 hr (8 days) for citric acid production.

In addition, our result in agreement with the most recent study by **Auta *et al.,* (2014)** where the quantity of citric acid produced varies with the type of tested microorganism. The rate of citric acid biosynthesis was studied and the maximum yield of citric acid (0.61 g/L) was after 5 days of fermentation. Extension of the fermentation period brought about depletion in the yield of citric acid produced. In batch fermentation of citric acid, the production started after a lag phase of one day and reached maximum at the onset of stationary phase or later. It might be due to the decreased available nitrogen in fermentation medium, the age of fungi, and depletion of sugar contents.

These explanations were also in corroboration with earlier reports by **van Suijdam *et al.,* (1980)** and **Chaudary *et al.,* (1978)** stated that much time causes the decrease of nitrogen and sugars in the substrate, thereby a reduction in citric acid production. This implies that the fermentation period used was suitable for the production of citric acid as higher yield is harvested at early stage thereby cutting down the cost needed to maintain the fermentation for longer time.

The local fungal strains, *A. niger* KSUD and FQW which gave the maximum amounts of citric acid at the 5th day, therefore, this incubation period was applied in the further experiments.

Table (4.II.2): Time course of citric acid production in shake submerged culture of *Aspergillus niger* Strains. Data represented as mean ± standard error.

A. *niger* Strains	Time (days)	pH	Growth (DW g/50 ml)	Citric Acid (mg/10ml)
FQW	0	6.43^b ±0.03	0.06^e ±0.00	3.32^d ±0.16
	1	5.95^a ±0.05	0.01^e ±0.02	3.00^d ±0.07
	2	4.04^c ±0.08	0.20^d ±0.01	5.21^c ±0.33
	3	3.61^d ±0.00	0.26^c ±0.01	9.57^b ±0.43
	4	3.37^e ±0.04	0.25^c ±0.035	9.19^b ±0.97
	5	2.54^f ±0.01	0.31^b ±0.01	13.19^a ±0.51
	6	2.29^g ±0.06	0.40^a ±0.00	14.09^a ±0.53
	LSD $_{0.05}$	0.13959	0.04538	1.5396
KSUD	0	6.50^a ±0.00	0.03^e ±0.00	4.54^c ±0.58
	1	6.12^b ±0.16	0.06^e ±0.00	5.70^c ±0.71
	2	4.12^c ±0.11	0.23^c ±0.02	6.59^c ±0.13
	3	3.56^d ±0.06	0.26^{cd} ±0.01	9.34^b ±0.66
	4	3.32^{de} ±0.04	0.40^{ab} ±0.03	11.77^b ±1.29
	5	3.02^e ±0.19	0.30^d ±0.02	16.91^a ±0.06
	6	2.43^f ±0.05	0.36^a ±0.00	17.16^a ±1.55
	LSD $_{0.05}$	0.3244	0.04712	2.64969
EMCC132	0	6.07^a ±0.00	0.08^e ±0.00	2.00^e ±0.00
	1	5.32^b ±0.02	0.12^e ±0.01	2.96^e ±0.32
	2	4.13^c ±0.05	0.22^d ±0.01	8.96^d ±0.71
	3	3.90^{cd} ±0.10	0.25^{cd} ±0.03	15.18^c ±0.92
	4	3.77^d ±0.11	0.28^c ±0.02	16.96^c ±0.08
	5	2.98^e ±0.03	0.33^b ±0.02	22.15^b ±1.20
	6	2.73^e ±0.02	0.38^a ±0.01	25.51^a ±0.87
	LSD $_{0.05}$	0.266	0.04909	2.1889

In Czapek-Dox broth medium. Initial pH was 6.5. Inoculum density was ~1.0X10^6 spore/ml. Cultures were grown with shaking at 100 rpm at 28°C.
Means in the same column followed by the same letter are not significantly different based on LSD at p = 0.05 according to Duncan's multiple range test.

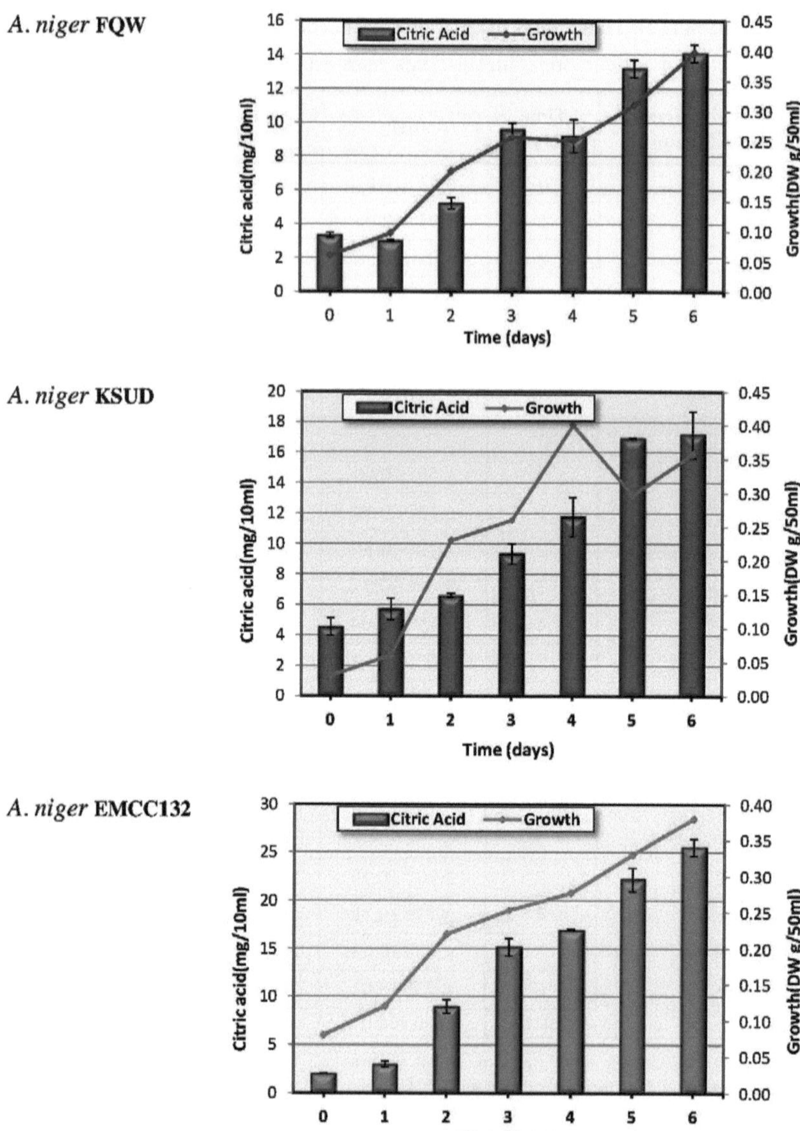

Fig. (4.II.2): Time course of citric acid production in shake submerged culture of *Aspergillus niger* Strains.

4. 3. Third Part: Optimization of Fermentation Condition for Citric Acid Production

In developing a fermentation process for obtaining a valuable product, the optimization of cultivation conditions and selection of appropriate substrates and the most favorable of their concentrations have primary importance due to their impact on the economy and feasibility of the process. Many studies have been intended to obtain high-yielded production using optimization processes with low cost as possible.

The improvement strategies used in citric acid production by *Aspergillus niger* were generally interested in classical optimization methods, and there are many reports about classical optimization methods for citric acid production based on varying one-factor-at-a time (**Ambati and Ayyanna 2001; Adham 2002; Barrington and Kim 2008; Kishore, 2008; Kishore and Reddy 2012; Maharani *et al.*, 2014**).

Citric acid productivity can be increased by optimizing its fermentation variables. For this purpose, studying the impacts of individual environmental and nutritional parameters on fungal growth and acids synthesis was performed as a basis for developing citric acid production process. This part was designed to find out appropriate conditions for high yield and thereby making it possible to manipulate citric acid synthesis in the two most local producers.

Citric acid production by *A. niger* is influenced by number of culture parameters. There are considerable variations in fermentation environments reported in the previous studies for citric acid production by *A. niger*. For achieving high production of citric acid, it is essential that the study of influence of physical and chemical environments on citric acid production (**Jianlong and Ping 1998**).

In order to optimize conditions for citric acid production by *A. niger* strains the following series of experiments include investigations on the influence of incubation temperature, starting reaction (H^+ conc. or pH) of the culture medium, an inoculums density as fungal spore, the static and shaken culture techniques (through agitation), carbon source, carbon source concentration, nitrogen source, nitrogen source concentration, phosphorous source and addition of alcohol as well as alcohol concentration, on the production of citric acid in a single batch culture. The aforementioned *A. niger* isolates were chosen after performing an initial screening of 98 isolates for their capability to produce citric acid. Then, according to the previous screening data, two strains were chosen for the further experiments depending on superior for citric acid production in addition to the reference strain. The tested strains were *Aspergillus niger* **FQW**, *Aspergillus niger* **KSUD** and *Aspergillus niger* EMCC132.

Optimization of various parameters was carried out with "One at a time" strategy keeping all other variable constant except one. The variable under study was varied and its optimal level was used for further investigation.

4.3.1. Effect of incubation temperature

Temperature is known to influence the metabolic rate of the organism involved in the process, which in turn determines the amount of the end product. Therefore, the optimum temperature for citric acid production was defined by assaying the citric acid amounts at different incubation temperatures ranging from 12 to 44°C, after 5 days as a favorite parameter refereeing to the pervious experiment results.

Data represented in Table (4.III.1) and illustrated in the Fig. (4.III.1) show a significant increment in citric acid production with the increase of incubation temperature up to 28°C for the three tested fungal strains (*A. niger* FQW, KSUD and EMCC132) and thereafter, higher temperature resulted in a gradually lowering of citric acid formation.

Maximum citric acid production by tested fungal strains, *A. niger* FQW, KSUD and EMCC132 were reached to 11.26, 15.25 and 22.02 mg/10ml, respectively at 28°C.

Obtained results indicated also that changing in pH values was accompanied by a noticeable of the yield of citric acid.

Moreover, growth of fungal strains were significantly lowered at both 12 and 44°C, but changing in fungal growth were insignificantly different in the temperature range 20-36°C (Table 4.III.1 and Fig. 4.III.1).

In general, temperature obviously influences the metabolic activity of cells. Where, the obtained results for growth of fungi at various incubation temperatures reveal that all tested *A. niger* strains could not tolerate higher temperature, that was appeared in the dry weight of the fungal biomass. Incubation temperature at 28°C was considered as an optimum cultivation temperature for the growth.

Temperatures between 25-30°C are usually considered optimal for fungi. However, a temperature above 35°C inhibits mycelial growth and favours oxalic accumulation. There is an exception for the thermo tolerant *A. niger* cultures which can produce citric acid even at 29-40°C (**Sanjay and Sharma 1994**).

In addition, **Vergano *et al.,* (1996)** reported that the temperature of fermentation medium is one of the critical factors that have a profound effect on the production of citric acid. Temperatures lower than 27°C slowed down growth and production substantially. When the temperature of medium was low, the enzyme activity was also low, giving no impact on the citric acid production. However when the temperature of medium was increased above 30°C, the biosynthesis of citric acid was decreased.

Table (4.III.1): Effect of incubation temperature on citric acid production in shake flask culture of *Aspergillus niger* strains. Data represented as mean ± standard error.

A. niger Strains	Incubation Temperature (°C)	pH	Growth (DW g/50 ml)	Citric Acid (mg/10ml)
FQW	12	6.45^a ±0.02	0.19^b ±0.02	4.00^c ±0.43
	20	2.73^d ±0.02	0.31^a ±0.01	7.01^b ±0.10
	28	2.39^d ±0.01	0.31^a ±0.04	11.26^a ±0.01
	36	3.79^c ±0.09	0.30^a ±0.03	7.86^b ±0.87
	44	5.12^b ±0.49	0.20^b ±0.02	3.22^c ±0.01
	LSD $_{0.05}$	0.6951	0.0750	1.3719
KSUD	12	6.44^a ±0.03	0.20^b ±0.03	7.04^c ±0.73
	20	2.71^d ±0.01	0.29^a ±0.01	6.52^{cd} ±0.09
	28	2.19^e ±0.00	0.30^a ±0.02	15.25^a ±0.08
	36	3.67^c ±0.00	0.32^a ±0.01	11.19^b ±0.35
	44	4.76^b ±0.07	0.16^b±0.01	5.10^d ±0.64
	LSD $_{0.05}$	0.1121	0.0507	1.4574
EMCC132	12	6.49^a ±0.00	0.21^b ±0.00	6.33^d ±0.11
	20	3.4^{cd} ±0.10	0.39^a ±0.01	8.55^c ±0.53
	28	3.08^d ±0.03	0.35^a ±0.015	22.02^a ±0.14
	36	3.70^c ±0.14	0.40^a ±0.06	20.90^b ±0.06
	44	4.63^b ±0.53	0.21^b ±0.02	5.09^e ±0.02
	LSD $_{0.05}$	0.4990	0.0953	0.7894

In Czapek-Dox broth medium. Initial pH was 6.5. Inoculum density was ~1.0×10^6 spore/ml. Cultures were grown with shaking at 100 rpm for 5 days.
Means in the same column followed by the same letter are not significantly different based on LSD at p = 0.05 according to Duncan's multiple range test.

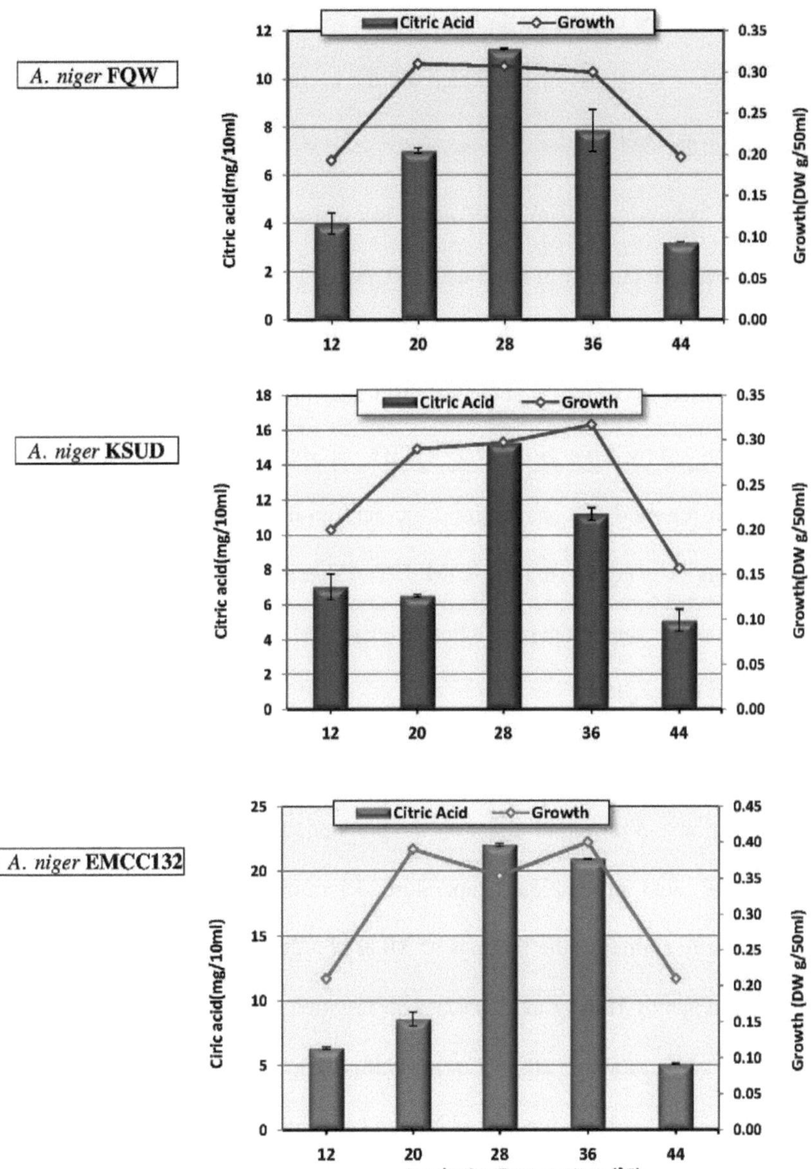

Fig. (4.III.1): Effect of incubation temperature on citric acid production in shake flask culture of *Aspergillus niger* strains.

It might be due to the fact that higher temperatures cause denaturation of enzyme citrate synthase, which aid the accumulation of other by-products such as oxalic acid.

Moreover, *Aspergillus niger* has grown well at 30°C and produced maximum amount of citric acid (80.68 g l^{-1}) followed by 20 °C with 74.21 g l^{-1} of citric acid (**Maharani, *et al.,* 2014**). Also, **Ali *et al.,* (2002)** have reported maximum amount of citric acid production (99.56 ± 3.59 g l^{-1}) achieved by *Aspergillus niger* GCBT7 at 30°C. **Kishore *et al.,* (2008)** has also reported that maximum citric acid production was recorded at 30°C using *Aspergillus niger* NCIM 705. The effect of temperature plays an important role in the production of citric acid by *Aspergillus niger* (**Grewal and Kalra, 1999**).

The temperature of fermentation medium is one of the critical factors that have a profound effect on the production of citric acid. The optimum citric acid production temperature determined in this study (28°C) is quietly similar to the findings of **Ali *et al.,* (2002)** and is also similar to the findings of **Haq *et al.,* (2002)** who reported that increment of incubation temperature above 30-35°C was inhibitory to citric acid production due to the increased production of by-product acids and inhibition of culture growth.

In addition different workers, agreed with the obtained findings, have also used 30°C as the cultivation temperature and maximum production of citric acid was obtained also when temperature of the medium was at 30°C, out of them **Prasad *et al.,* (2014); El-Hussein *et al.,* (2009) and Ali *et al.,* (2002)** reported that a temperature of 30°C was the best for citric acid fermentation by *A. niger* strains. When the temperature of medium was low, the enzyme activity was also low (at 24°C), giving no impact on the citric acid production. But when the temperature of medium was increased above 30°C, the biosynthesis of citric acid was decreased (at 36°C). It might be due to the accumulation of by-products such as oxalic acid. In addition, **Kishore (2008)** and **Kareem *et al.,* (2010)** recorded optimum temperature of 30°C for maximum citric acid production using *A. niger* strains and **Karthikeyan and SivaKumar (2010)** who also recorded 28°C as optimum temperature for citric acid production by *A. niger* using banana peels as substrate.

Similarly, the obtained results in agreement with the findings of **Kareem and Rahman (2013)** who stated that the temperature of fermentation medium is one of the critical factors that have a profound effect on the production of citric acid. When the temperature of medium was low, the enzyme activity was also low, giving no impact on the citric acid production. But when the temperature of medium was increased above

55ºC, the biosynthesis of citric acid decreased. It might be due to the accumulation of byproducts and inhibition culture growth.

In addition, higher than optimal temperatures result in enzyme denaturation and inhibition, excess moisture losses and growth arrest while lower temperatures lead to lower metabolic activity (**Adinarayana *et al.*, 2003**).

The effect of different temperatures on citric acid production by *A. niger* isolate was studied. A temperature of 55∘C was found optimal for citric acid production as maximum citric acid of 0.62 g/L was produced at this temperature by *A. niger* isolate. Further increase in temperature gradually decreased citric acid productivity by the isolate. The temperature of fermentation medium is one of the critical factors that have a profound effect on the production of citric acid (**Auta *et al.*, 2014**).

According to obtained data in our experiments, incubation temperature of 28°C was selected for further experiments for improvement of citric acid production.

4.3.2. Effect of initial pH

The metabolic activity of fungus is very sensitive to pH level of media. The initial pH of the medium was found to have an impact on citric acid production by *A. niger* (**Kim *et al.,* 2004**). Most filamentous fungi are observed to grow well under slightly acidic conditions, ranging from 3 to 6, but some fungi are able to growth at a pH value below 2 to better compete against bacteria (**Fawole and Odunfa 2003**).

The favourable initial pH is one of the most important steps for the successful progression of citric acid fermentation. The objective of this experiment was to evaluate the effect of the initial pH of the nutrient solution on citric acid production by *A. niger* local strains and the reference strain.

The influence of initial pH value of the culture medium on growth and citric acid production by *Aspergillus niger* strains (FQW, KSUD and *A. niger* EMCC132) was investigated at initial pH values 3, 5, 6.5, 8 and 10. The initial pH values were adjusted by diluted acid or alkali. The flask cultures were incubated on rotary shaker for five days at optimal temperature 28°C attained in the foregoing experiments.

The results on the effects of different initial pH values on citric acid production, mycelial dry weight and changing in cultures' pH by *A. niger* strains are recorded in Table (4.III.2) and illustrated by Figure (4.III.2). The obtained results indicated that maximum citric acid concentrations by *A. niger* FQW, KSUD and EMCC132 (11.26, 15.25 and 22.27 mg/10 ml, respectively), was obtained at pH 6.5.

Moreover, a very sharp decrease in citric acid yield was observed when the initial pH values were increased to 8.0 and to 10.0. This reduction was also significantly noticed and accompanied by a similar reduction in mycelial dry weight.

The effect of the initial pH of nutrient solution on citric acid production was evaluated and concluded that the initial pH had a considerable effect on the growth of *A. niger* strains and production of citric acid. From the results of our work, it could be concluded that citric acid production and growth yield were strongly affected by initial pH level and apparently the maintenance of a favorable initial pH 6.5 is essential for successful production of citric acid in further experiments.

In general, initial pH value of the culture medium mainly controls the fungal growth and citric acid production as well. Different pH optima have been report by several researchers.

Table (4.III.2): Effect of initial pH value on citric acid production in shake flask culture of *Aspergillus niger* strains. Data represented as mean ± standard error.

A. *niger* Strains	Initial pH	pH	Growth (DW g/50 ml)	Citric Acid (mg/10ml)
FQW	3.0	2.39^b ±0.02	0.47^a ±0.04	6.12^b ±0.02
	5.0	2.46^a ±0.01	0.43^a ±0.01	6.32^b ±0.30
	6.5	2.39^b ±0.01	0.34^b ±0.02	11.26^a ±0.01
	8.0	2.14^c ±0.01	0.24^c ±0.04	6.59^b ±0.77
	10.0	2.34^b ±0.03	$0.32^{b,c}$±0.03	6.15^b ±0.49
	LSD $_{0.05}$	0.0579	0.0920	1.3539
KSUD	3.0	2.41^b ± 0.04	0.26^{bc}±0.01	9.31^c ±0.77
	5.0	2.54^a ± 0.06	0.49^a ±0.03	11.31^b ± 0.35
	6.5	2.18^c ± 0.01	0.31^b ±0.01	15.25^a ±0.08
	8.0	2.07^d ± 0.01	0.26^{bc}±0.01	8.24^{cd} ±0.29
	10.0	2.12^{cd} ± 0.01	0.24^c ±0.01	7.14^d±0.05
	LSD $_{0.05}$	0.1062	0.0553	1.2619
EMCC132	3.0	3.44^a ±0.07	0.33^{ab} ±0.01	8.72^d ±0.52
	5.0	3.22^{ab}±0.07	0.29^{bc} ±0.03	15.93^b ±0.51
	6.5	3.09^b ±0.29	0.37^a ±0.01	22.27^a ±0.01
	8.0	2.55^c ±0.12	0.26^c ±0.00	11.15^c ±0.03
	10.0	3.00^b ±0.04	0.3^{bc} ±0.00	8.38^d ±0.01
	LSD $_{0.05}$	0.2257	0.0443	1.025

In Czapek-Dox broth medium. Inoculum density was ~1.0×10^6 spore/ml. Cultures were grown with shaking at 100 rpm at 28°C for 5 days.
Means in the same column followed by the same letter are not significantly different based on LSD at p = 0.05 according to Duncan's multiple range test.

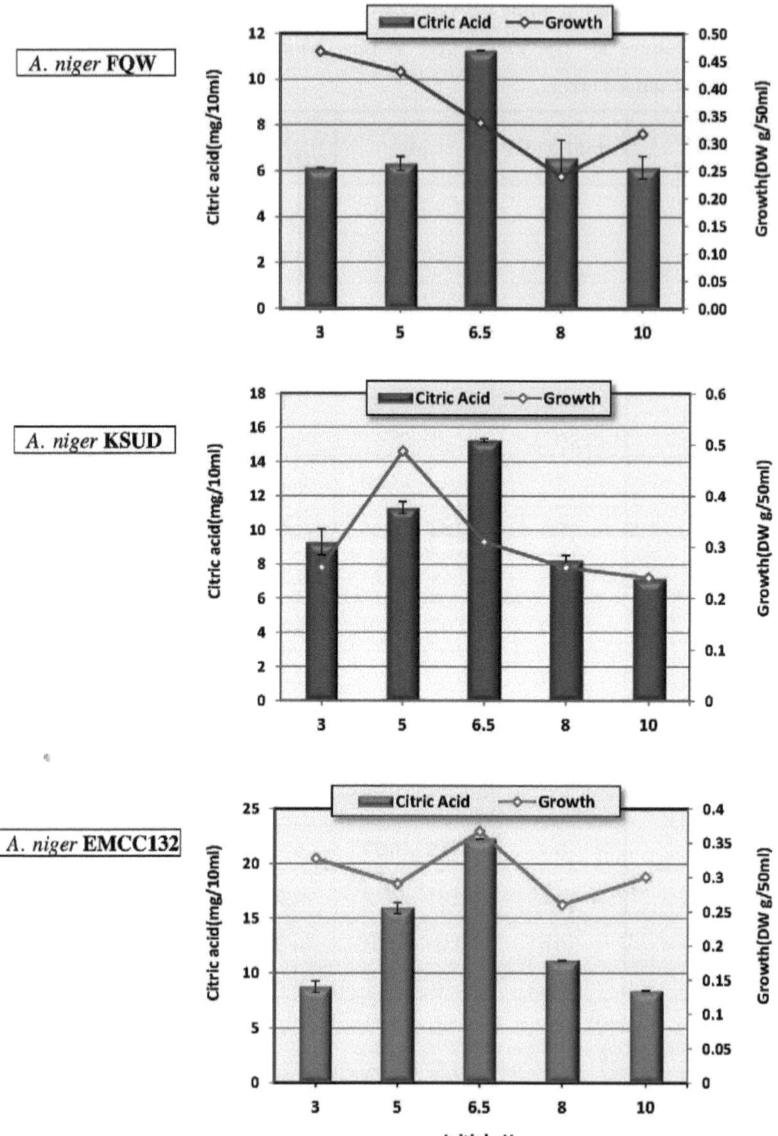

Fig. (4.III.2): Effect of initial pH value on citric acid production in shake flask culture of *Aspergillus niger* strains.

The obtained results are in line with **Pessoa *et al.*, (1982 and 1984)** studied the influence of low and high initial pH values and reported that decrease in initial pH value lower than 3.5 results in reduction in citric acid production. It might be due to that at low pH, the ferrocyanide ions were more toxic for the growth of mycelium. A higher initial pH leads to the accumulation of oxalic acid. Also **Paul *et al.*, (1999)** mentioned that Oxalic acid is an undesirable by-product of citric acid fermentation and is considered to be a main impurity in the commercial citric acid process from *A. niger* at higher initial pH.

In addition, **Ali *et al.*, (2002)** insured that the maintenance of a favorable pH is very essential for the successful production of citric acid. Effect of different pH (4.5-7.0) on the citric acid production was studied and the highest value of citric acid yield (96.12 ± 3.5 g/l) when initial pH of the fermentation medium was kept at 6.0 by culture of *Aspergillus niger* GCBT7. A lower initial pH was reported to inhibit the growth of *A. niger* and is thus expected to negatively affect citric acid production.

Similar results were explained in previous studies. Where the effect of different initial pH (4.5 - 7.0) of the fermentation media was also studied by **El-Aasar (2006)** who found that maximum citric acid (47.63 g/l) were obtained when the initial pH of the fermentation medium was adjusted to 5.5. Decrease in pH caused reduction in both protein and citric acid yield.

Also, **Prasad *et al.,* (2014)** mentioned that the production of citric acid was studied at different pH (4.8 to 6.0) levels. Maximum yield of citric acid was obtained (9.3g/100 ml) when the pH was maintained at 5.4 with the *A. niger*.

The maintenance of a favorable pH is very essential factor for the successful production of citric acid, pH 5 was the optimum. Decrease in pH caused reduction in citric acid production. It might be due to that at low pH, the ferrocyanide ions were more toxic for the growth of mycelium (**Vasanthabharathi et al., 2013**).

It is likely that at low pH, the ferrocyanide ions were more toxic for the growth of mycelium and consequentially citric acid production. While higher initial pH leads to the accumulation of oxalic acid (**Pessoa et al., 1982**; **Ali *et al.,* 2002**; **El-Holi and Al-Delaimy 2003**).

On the other hand, **Shadafza *et al.,* (1976)** reported that a pH higher than 3.5 leads to accumulation of oxalic acid and gluconic acid at the expense of citric acid production. Also **Ali *et al.,* (2005)** reported the maximum concentration of citric acid production (140 g/l) by *A. niger* was observed at the pH 3.5. As well **Maharani *et al.,* (2014)** found that the *Aspergillus niger* produced higher concentration (82.65 g 1-1) of citric acid at the pH 3.5. This is due to the evidence of strong acidic conditions that are required for *Aspergillus niger* growth and production of citric acid.

Mostly, the fungal strains are seemed to be grown well under acidic conditions ranging from 3 to 6 (**Fawole and Odunfa, 2003**).

The utilization of *Parkia biglobosa* fruit pulp as substrate for citric acid production by *Aspergillus niger* was evaluated and the effect of initial pH on citric acid production was investigated. The result showed that increase in pH brought about decrease in citric acid production. Maximum yield of 1.15 g/L was obtained at pH 2. Further increase in pH was found to decrease citric acid production (**Auta *et al.,* 2014**).

The pH of a culture may change in response to microbial metabolic activities. The most obvious reason is the secretion of organic acids, such as citric acid, which will cause the pH decrease. Changes in pH kinetics also depend highly on the microorganism. With *Aspergillus* sp., *Penicillium* sp. and *Rhizopus* sp., pH can drop very quickly to less than 3.0. For other groups of fungi such as *Trichoderma, Sporotrichum, Pleurotus,* pH is more stable between 4.0 and 5.0. The nature of the substrate and production technique is also influence pH kinetics. In this way initial pH must be very well defined and optimized depending on the microorganism, substrate and production technique (**Soccol *et al.,* 2006**).

The pH of the medium is important at two different times in the fermentation for different reasons. Firstly, the spores require a pH 5 in order to germinate. All fermentations are started from spores even if the culture is "scaled up" through seed fermenters before being transferred to production fermenters. Secondly, the pH for citric acid production needs to be low (pH $\leq$ 2). A low pH reduces the risk of contamination of the fermentation with other microorganisms. A low pH also inhibits the production of unwanted organic acids (gluconic acid, oxalic acid) and this makes the recovery of citric acid from the broth simpler. The uptake of ammonia by the germinating spores causes a corresponding release of protons that lower the pH to approximately the right level after spore germination has occurred. Increasing the pH to 4.5 during the production phase reduces the final yield of citric acid by up to 80% (**Papagianni, *et al.,* 1999**).

Culture pH can have a profound effect on citric acid production by *A. niger* because certain enzymes within the TCA cycle are pH sensitive. Maintenance of a low pH during fermentation is vital for a good yield of citric acid and it is generally considered necessary for the pH to fall to around 2.0 within a few hours of the initiation of the fermentation. Failing this, the yields are reduced (**Mattey, 1992**).

4.3.3. Effect of inoculums density

Spore inoculum level is another parameter that influences citric acid fermentation and improvement. The objective of the present experiment was to evaluate the effect of the inoculums density on citric acid production by *A. niger* local strains and the reference strain.

The influence of inoculums density on fungal growth and citric acid production by *Aspergillus niger* strains (FQW, KSUD and *A. niger* EMCC132) was investigated. The conidial density (from 5 days old cultures) of the used inoculums were (1.0, 2.0 and 3.0) x10^6 spores/ml. The flask cultures were incubated on rotary shaker. The initial pH, incubation temperature and incubation period were maintained at optimal levels that attained in the foregoing experiments.

Data present in Table (4.III.3) and Figure (4.III.3) obviously indicate that citric acid yields by the three strains of *A. niger* were significantly increased by increment of the conidial density in the inoculums after 5 days of flask fermentation at 28°C.

Among different density of used inoculums, citric acid productivity was proportionally increased by conidial density increasing. Where citric acid yields were reached to the highest values with 3.0X10^6 inoculums size of all tested *A. niger* strains. Changing in pH values and fungal growth of *A. niger* FQW, KSUD and EMCC132 after 5 days of shaking flask cultures were not significantly appeared by increasing the inoculums density (as shown in Table 4.III.3).

According to obtained data, inoculums size 3.0×10^6 spores/ml was selected for further experiments for improvement of citric acid production.

In agreement with our findings of inoculums density between 1×10^4 to 1×10^9 spores/ml was found to be suitable for citric acid production by *Aspergillus niger* in submerged fermentation (**Favela-Torres *et al.*, 1998; Adham 2002; Papagianni and Mattey 2006**).

Table (4.III.3): Effect of inoculums density on citric production in shake flask culture of *Aspergillus niger* strains. Data represented as mean ± standard error.

A. *niger* Strains	Inoculum Density (X10^{6}spore/ml)	pH	Growth (DW g/50 ml)	Citric Acid (mg/10ml)
FQW	1.0	2.65^a ± 0.04	0.31^a ±0.05	10.89^b ±0.34
	1.5	2.55^{ab} ±0.05	0.39^a ±0.02	11.82^b ±1.34
	2.0	2.47^b ±0.01	0.35^a ±0.02	16.97^a ±0.04
	LSD $_{0.05}$	0.1274	0.1130	2.755
KSUD	1.0	2.66^a ±0.09	0.55^{ab} ±0.00	10.19^c ±0.03
	1.5	2.85^a ±0.10	0.44^{bc} ±0.07	14.16^b ±0.06
	2.0	2.59^a ±0.00	0.59^a ±0.00	18.62^a ±0.11
	LSD $_{0.05}$	0.2778	0.1385	0.2507
EMCC132	1.0	3.04^a ± 0.01	0.33^{ab} ±0.05	14.35^b ±0.11
	1.5	3.04^a ±0.10	0.38^a ±0.03	21.53^a ±0.04
	2.0	2.94^a ±0.01	0.23^b ±0.01	23.02^a ±3.37
	LSD $_{0.05}$	0.1976	0.1209	6.7424

In Czapek-Dox broth medium. Initial pH was 6.5. Cultures were grown with shaking at 100 rpm at 28°C for 5 days.

Means in the same column followed by the same letter are not significantly different based on LSD at p = 0.05 according to Duncan's multiple range test.

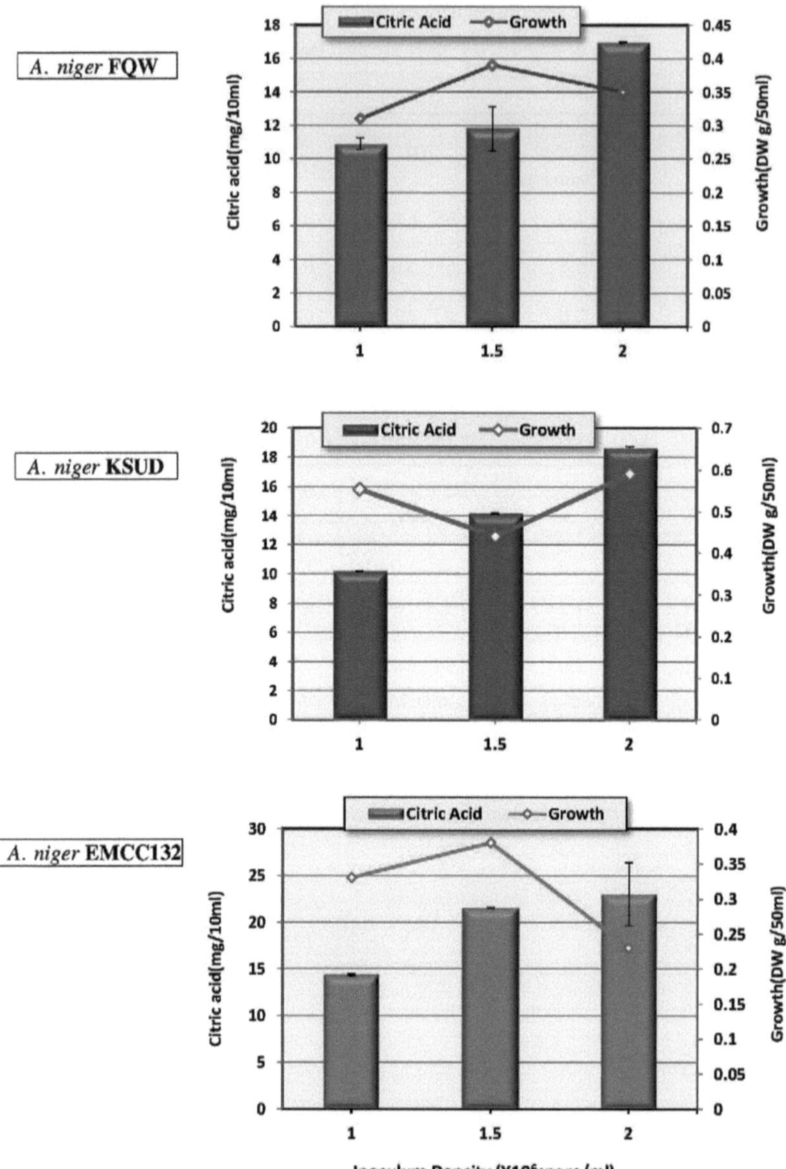

Fig.(4.III.3): Effect of inoculums density on citric production in shake flask culture of *Aspergillus niger* strains.

According to the previous literatures, citric acid production is known to be affected by inoculums density. **Van-Suijdam *et al.*, (1980)** reported that 1.0% vegetative inoculums to be adequate for optimal production of citric acid. The citric acid yield was lower at 5% of the vegetative inoculum. This may be due to the clumping of *Aspergillus niger* at higher inoculum concentration. The increase in the number of inoculated spores primarily increases the level of acid production; however, on the long run this increase tends to force the process toward cellular crowding and to favor the consumption of sugar for biomass reproduction, thus resulting in a reduction in production of citric acid. The excessive reduction in the level of inoculation also causes prolongation in the adaptation phase, therefore decreasing an ideal outcome. This result suggests that there appears to be an optimum amount of inoculum required for citric acid production. This disagrees with the findings of **Ali (2004)** who reported that the 5% conc. of inoculum was optimum for maximum citric acid production (96.55 g l^{-1}) using *Aspergillus niger* under submerged fermentation condition. The 5% conc. of the inoculums is seems to be optimum for citric acid production.

A high inoculum density leads to population over-crowding, higher nutrient competition and rapid exhaustion of nutrients. Up to a specific limit, metabolite production generally increases with inoculum density (**Nampoothiri *et al.*, 2004**). At the lower inoculum density, metabolite production drops and contamination risks increase due to an insufficient cell population.

In addition, **Kirimura** and his colleagues **(2011)** suggested that the highest level of citric acid is produced when the age of spores is less than seven days, because older spores tend to consume the initially produced acid before the fermentation process is completed.

Otherwise, the increase in the number of inoculated spores primarily increases the level of acid production, however, in the long run this increase tends to force the process toward cellular crowding and to favor the consumption of sugar for biomass reproduction, thus resulting in a reduction in production of citric acid. The excessive reduction in the level of inoculation also causes prolongation in the adaptation phase, therefore decreasing an ideal outcome (**Alam *et al.,* 2011 and Dhillon *et al.,* 2013**).

When **Auta *et al.,* (2014)** studied the effect of vegetative inoculum size (1–5%) on citric acid production by *Aspergillus niger*. They mentioned that the maximum citric acid production of 0.53 g/L was obtained with 3% inoculum size. As the inoculum size increased, citric acid production decreased.

4.3.4. Effect of Agitation

In submerged fermentations, agitation is necessary for good mass. Agitation can affect filamentous micro-organisms in various ways, causing rate variations of growth and product formation, development of different morphological forms and breakage of filaments. Since filamentous micro-organisms are of great industrial importance, the relationship between mechanical parameters, morphology and productivity has attracted the attention of many investigators (**Papagianni** *et al.,* **1994**).

Particular morphological forms, often agitation induced, have been considered as more suitable for certain fermentations. Several authors have noted shorter and wider cells in fungal filaments, with greater branching frequencies and reduced aggregate sizes under conditions of high agitation (**Mitard and Riba 1988**).

Studying the behavior of three citric acid producing *Aspergillus niger* strains, **Ujcova** *et al.,* **(1980) and Papagianni** *et al.,* **(1998)** observed that higher speeds resulted in thicker, densely branched filaments, while production was optimum within a narrow range of speeds and a drop in productivity resulted from higher stirrer speeds.

As well, agitation speed is a very important factor in the fermentation process since it will increase the amount of dissolved oxygen in the cultivation medium and it also exerts influence on the availability of other nutrients in the medium. Agitation has direct influence on the amount and rate of citric acid production. The proper agitation is important for the maintenance of a suitable oxygen supply to the mould growth (**Sakurai *et al.*, 1996 and Papagianni *et al.*, 1998**).

Thus, the effect of aeration on growth and citric acid production was studied. The obtained results are presented in Table (4.III.4) and illustrated in Figures (4.III.4a & 4.III.4b). Aeration by reciprocal shaking at a speed of 100 rpm resulted in greater citric acid production for all tested *A. niger* strains. After 5 days of flask cultures, citric acid was within range 7.72-12.38 mg/10 ml by *A. niger* FQW, *A. niger* KSUD and *A. niger* EMCC132, respectively, in the case of static cultures. However, citric acid amounts reached to 16.97-23.02 mg/10 ml in the shaken broth cultures of *A. niger* FQW, *A. niger* KSUD and *A. niger* EMCC132, respectively.

The results also show that the type of culturing had a pronounced effect on biomass dry matter (Table 4.III.4 and Fig. 4.III.4b). All strains obviously possessed high growth in static than in shaken cultures, where fungal biomass obtained after 5 days in static cultures were nearly ten times higher, except in case of *A. niger* EMCC132 strain was not pronounced effect (Table 4.III.4 and Fig. 4.III.4b).

It was also found that in either static or shaken cultures the final pH values of the citric acid production medium lied in acidic range (2.47-3.16). In any case, the type of culturing had little effect on the final pH value of the citric acid production culture, which varied within a very limited range with the type of culturing. Where, slightly acid was produced in shaken cultures (Table 4.III.4).

For that shaken conditions were selected for further investigations on the tested strains, *A. niger* FQW, KSUD and EMCC132 for improving their citric acid production.

Agitation is necessary both for aeration of cultures and citric acid production, similar kinds of findings have also been reported by **Papagianni *et al.*, (1998)** who demonstrated citric acid production of *A. niger* PM1 were agitation dependent and closely linked.

Increased aeration rates led to enhanced yields and reduced fermentation time. Agitation is an additional reason why small compact pellets are the preferred mycelial forms of *A. niger* during fermentation **(Kubicek and Rohr 1986)**. The influence of dissolved oxygen concentration on citric acid formation has been examined by **Grewal and**

Kalra (1995). Also, intensive agitation is associated with development of short, thick, and highly branched filaments that overproduce citric acid **(Papagianni 2007)**.

Table (4.III.4): Effect of shaking on growth and citric acid production by tested fungal strains.

Fungal strains	pH		Growth (DW mg/50 ml)		Citric Acid (mg/10ml)	
	Static	Shaking	Static	Shaking	Static	Shaking
A. niger FQW	3.16	2.47	3.12	0.35	7.71	16.97
A. niger KSUD	3.21	2.59	3.21	0.59	8.12	18.62
A. niger EMCC132	2.65	2.94	0.20	0.23	12.38	23.02

In Czapek-Dox broth medium. Initial pH was 6.5. Inoculum density was ~2.0×10^6 spore/ml. Cultures were grown at 28°C for 5 days.

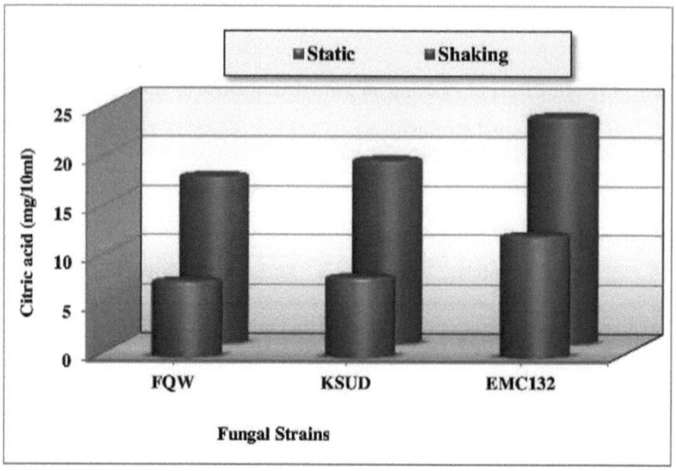

Table (4.III.4a): Effect of shaking on citric acid production by
tested fungal strains.

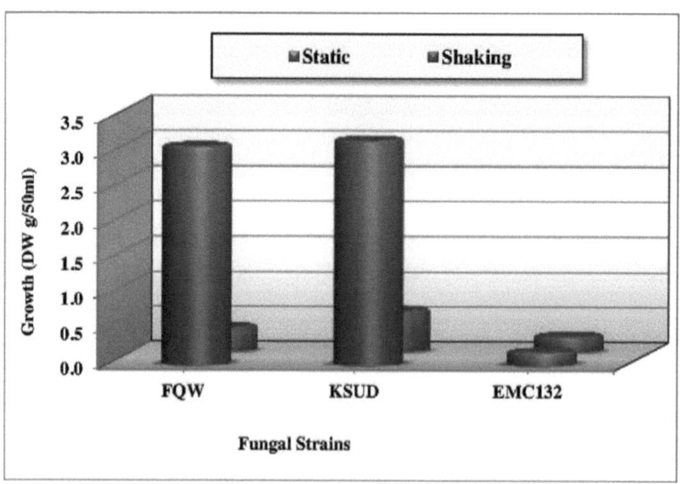

Table (4.III.4b): Effect of shaking on fungal growth by tested
fungal strains.

4.3.5. Effect of Carbon Sources

Citric acid accumulation is strongly influenced by the type of carbon source. The carbon source for the citric acid fermentation has been the focus of much study, frequently with a view to the utilizing polysaccharide sources. Citric acid accumulation is strongly affected by the nature of the carbon source. The presence of easily metabolized carbohydrates has been found essential for good production of citric acid (**Soccol *et al.*, 2006**).

In general, only sugars that are rapidly taken up by the fungus allow a high final yield of citric acid. Polysaccharides, unless hydrolyzed, are generally not a useful raw material for citric acid fermentation because they are broken down too slowly to match the high rate of sugar catabolism required for citric acid production. The slow hydrolysis of polysaccharides is due to the low activity of the hydrolytic enzymes at the low pH that is necessary for producing citric acid (**Mattey 1992**).

There is several factors affect citric acid production and among these, carbon source has been found to play an important role in citric acid biosynthesis. So that for enhancing the production of citric acid in the present study, *A. niger'* strains (*A. niger* FQW, KSUD and EMCC132) were cultivated on the Czapek-Dox broth medium amended with different carbon sources (sucrose, maltose, glucose, fructose, lactose and arabinose) in flask shack culture at 28°C for 5 days to search for inducers of the citric acid overproduction.

Data represented in Table (4.III.5) and Figure (4.III.5) indicated that maltose followed by sucrose have been employed as the preferred carbon sources for citric acid production by *A. niger* FQW (23.02 and 17.97 mg/10 ml, respectively) and *A. niger* KSUD (20.99 and 16.25 mg/10 ml, respectively) however in the case of reference strain *A. niger* EMCC132, maltose followed by glucose as the preferred carbon sources for citric acid production, 41.74 and 24.61 mg/10 ml, respectively.

All of the other used sugars showed lower citric acid yield. Moreover, arabinose and lactose had shown significantly an inhibitory effect on citric acid production. In general, maltose served as the best carbon source for citric acid production by all tested *A. niger'* strains (*A. niger* FQW, KSUD and EMCC132).

It is clear that citric acid production by fungal tested strains is dependent on nutritional conditions and especially on carbon source.

The fungal biomass was varied widely among the carbon sources utilized. Where, the highest biomass values were obtained in the presence

Table (4.III.5): Citric acid production by *Aspergillus niger* strains on various carbon sources in shake flask culture. Data represented as mean ± standard error.

A. *niger* Strains	Carbon Sources	pH	Growth (DW g/50 ml)	Citric Acid (mg/10ml)
FQW	Control	2.95^c ±0.05	0.35^b ±0.01	17.97^b ±0.29
	Maltose	2.59^d ±0.02	0.40^a ±0.01	23.02^a ±0.03
	Glucose	2.78^{cd} ±0.08	0.29^c ±0.00	9.96^c ±0.12
	Fructose	2.93^c ±0.07	0.19^e ±0.02	3.29^f ±0.04
	Lactose	3.60^b ±0.12	0.25^d ±0.01	3.91^e ±0.11
	Arabinose	4.24^a ±0.02	0.24^d ±0.00	5.22^d ±0.10
	LSD $_{0.05}$	0.2153	0.03248	0.4424
KSUD	Control	2.78^c ±0.01	0.30^{cd} ±0.00	16.25^b ±0.11
	Maltose	2.95^c ±0.11	0.52^a ±0.03	20.99^a ±0.27
	Glucose	2.77^c ±0.12	0.32^c ±0.01	10.70^c ±0.30
	Fructose	2.84^c ±0.07	0.40^b ±0.01	11.20^c ±0.43
	Lactose	4.51^a ±0.09	0.26^d ±0.01	5.30^d ±0.77
	Arabinose	3.64^b ±0.14	0.27^d ±0.02	5.97^d ±0.24
	LSD $_{0.05}$	0.30529	0.05066	1.2680
EMCC132	Control	2.94^c ±0.01	0.43^a ±0.01	23.31^c ±0.14
	Maltose	3.16^c ±0.19	0.31^b ±0.01	41.74^a ±0.38
	Glucose	3.43^c ±0.10	0.26^c ±0.00	24.61^b ±0.23
	Fructose	3.40^c ±0.06	0.26^d ±0.00	7.90^d ±0.07
	Lactose	5.15^b ±0.64	0.18^d ±0.01	5.89^e ±0.32
	Arabinose	5.10^a ±0.00	0.13^e ±0.01	3.22^f ±0.17
	LSD $_{0.05}$	0.8459	0.02965	0.7449

In modified Czapek-Dox broth medium containing sucrose as control. Initial pH was 6.5. Inoculum density was ~2.0X10^6 spore/ml. Cultures were grown with shaking at 100 rpm at 28°C for 5 days. Means in the same column followed by the same letter are not significantly different based on LSD at p = 0.05 according to Duncan's multiple range test.

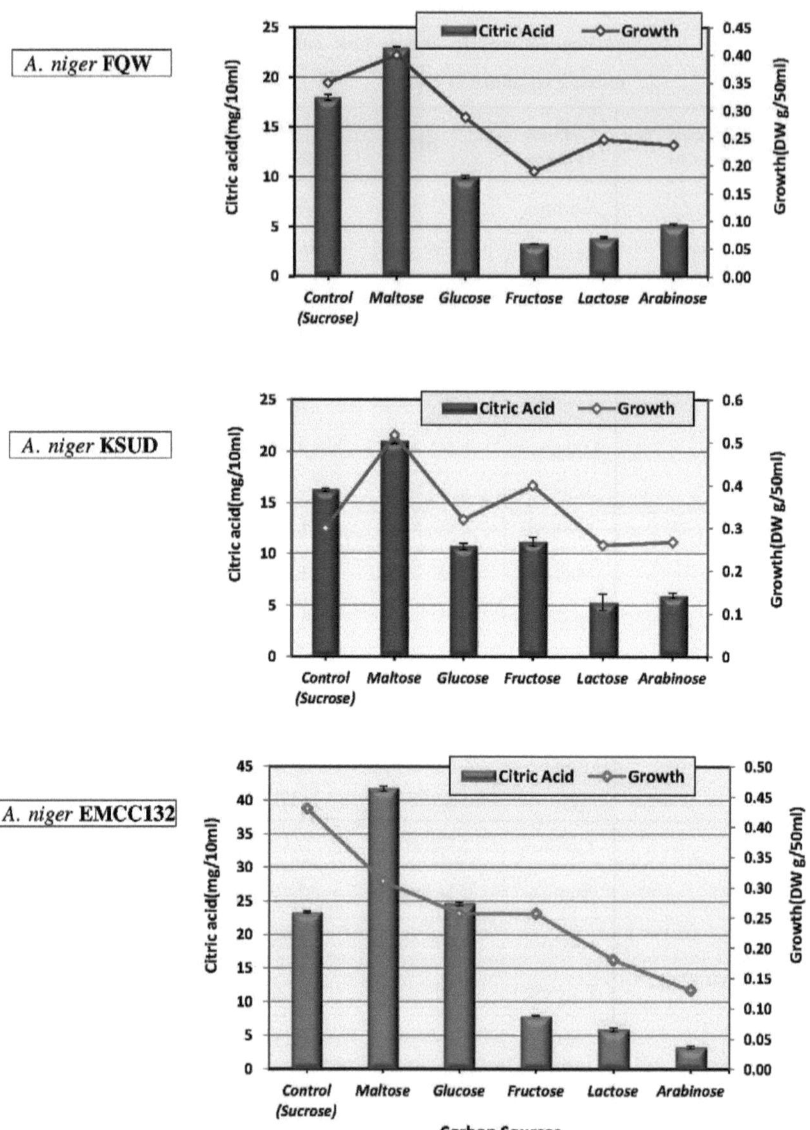

Table (4.III.5): Citric acid production by *Aspergillus niger* strains on various carbon sources in shake flask culture.

of maltose and sucrose, while the fungal growth were suppressed in the presence of lactose and arabinose that is reflect on the acid production.

Among the used carbon substrates (which are mostly sugars of hexoses, mannoses, mono- or disaccharide nature), the previous studies were varied. For instance, **Hossain *et al.*, (1984)** explained that the nature of the sugar source has marked effect on citric acid production by *A. niger*. He also showed that sucrose was the most favourable carbon source followed by glucose, fructose and galactose. Galactose contributed to a very low growth of fungi and did not favour citric acid accumulation.

Other sources of carbon such as sorbose, ethanol, cellulose, manitol, lactic, malic and a-acetoglutaric acid, allow a limited fungal growth and low citric acid production. Starch, pentoses (xyloses and arabinoses), sorbitol and pyruvic acid slow down growth, though the production is minimal (**Xu *et al.*, 1989**).

Sucrose is the traditional commercial substrate for citric production although glucose, fructose and maltose have also been used as substrates for citric acid production (**Xu *et al.*, 1989**).

Generally, the presence of carbohydrates which are rapidly taken up by microorganisms has been found essential for a good production of citric acid (**Yokoya 1992**).

The superiority of sucrose over glucose and fructose has been documented by **Gupta *et al.*, (1976)**.

Moreover, among the easily metabolized carbohydrates, sucrose is the most favourable carbon source followed by glucose, fructose and galactose for citric acid production (**Yokoya 1992; Dasgupta *et al.,* 1994; Vandenberghe 1999**).

However, it is also found that glucose is not suitable for higher production of citric acid (**Sukesh *et al.,* 2013**).

In contrary of the obtained results **Boominathan *et al.,* (2012)** found that maximum citric acid production was recorded in the glucose incorporated medium and the low amount of citric acid was recorded in the maltose supplemented medium.

Maltose sugar which gave the highest citric acid production was used in further experiments for optimization of other medium components.

4.3.6. Effect of Carbon Source Concentrations

Carbon compounds are the sources of carbon skeleton and energy of microorganism cell. The accumulation of citric acid is strongly influenced by the composition of the fermentation medium, particularly in the submerged process. Among factors that have been shown to exert an effect on citric acid fermentation, concentration of the carbon source. Where, citric acid accumulation is strongly influenced by the concentration of carbon source as reported by **Soccol *et al.,* (2006)**.

All of the previous studied parameters were utilized as optimum conditions that have been accessed and led to the increasing of citric acid production by *A. niger* local strains FQW and *A. niger* KSUD up to 23.02 and 20.99 mg/10 ml, respectively (Table 4.III.5).

The influence of different maltose concentrations (20, 25, 30, 35 and 40 g/l) on the production of citric acid production by all tested *A. niger'* strains (*A. niger* FQW, KSUD and EMCC132) was tested in flask cultures that adjusted to pH 6.5 and incubated in shaker incubator (100 rpm) at 28°C for 5 days.

The results represented in Table (4.III.6) and Figure (4.III.6) indicates that the citric acid yields proportional produced by carbon substrate concentrations increment, and the citric acid production was reached to the maximum values 25.55, 42.35 and 47.06 mg/10 ml by *A. niger* FQW, KSUD and EMCC132, respectively, at 40 g/l maltose concentration.

In addition, it was found that mycelial dry weight of all tested *A. niger* strains were significantly enhanced with the increase of initial sugar concentration.

Presented results concerning the favorable maltose concentration (4%) for maximal citric acid production, are in a range with certain previous researches.

Since (1960) **Kovats** mentioned that initial sugar concentration was critical for citric acid production and other organic acids produced by *Aspergillus niger*.

Pazouki *et al.,* **(2000)** who reported that a high sugar concentration leads to accumulation of greater amounts of residual sugars making the process uneconomical while a lower sugar concentration leads to lower yields due to accumulation of oxalic acid in the culture broth.

Moreover, **Hossain** *et al.,* **(1984)** suggested that high concentrations of appropriate carbon sources lead to repression of α-keto-glutarate dehydrogenase, hence explaining the effect of the sugar concentration and source in terms of enzyme repression.

Table (4.III.6): Effect of maltose concentrations on citric acid production in shake flask culture of *Aspergillus niger* Strains. Data represented as mean ± standard error.

A. *niger* Strains	Maltose Concen. (g/l)	pH	Growth (DW g/50 ml)	Citric Acid (mg/10ml)
FQW	20	3.76^a ±0.03	0.24^d ±0.02	5.71^c ±0.09
	25	3.72^a ±0.04	0.33^c ±0.02	5.38^c ±0.01
	30	3.61^a ±0.01	0.39^b ±0.00	23.77^b ±0.74
	35	3.58^a ±0.07	0.45^a ±0.02	23.22^b ±0.19
	40	3.52^a ±0.23	0.49^a ±0.01	25.55^a ±0.62
	LSD $_{0.05}$	0.3432	0.0506	1.3881
KSUD	20	2.51^c ±0.01	0.18^d ±0.01	6.00^d ±0.04
	25	2.56^{bc} ±0.00	0.23^c ±0.01	6.63^d ±0.28
	30	2.60^b ±0.009	0.25^c ±0.00	22.36^c ±1.05
	35	2.59^b ±0.03	0.28^b ±0.00	30.49^b ±0.28
	40	2.69^a±0.02	0.33^a ±0.02	42.35^a ±7.07
	LSD $_{0.05}$	0.0539	0.0244	10.0914
EMCC132	20	4.09^a ±0.06	0.18^d ±0.02	8.07^d ±0.10
	25	4.05^{ab}±0.02	0.26^c ±0.01	10.94^c ±0.06
	30	4.02^{ab}±0.02	0.30^b ±0.01	44.66^b ±0.86
	35	3.86^c ±0.00	0.33^b ±0.00	43.90^b ±1.02
	40	3.98^b ±0.01	0.41^a ±0.00	47.06^a ±0.22
	LSD $_{0.05}$	0.0968	0.0357	2.3705

In modified Czapek-Dox broth medium. Initial pH was 6.5. Inoculum density was ~2.0X10^6 spore/ml. Cultures were grown with shaking at 100 rpm at 28°C for 5 days.

Means in the same column followed by the same letter are not significantly different based on LSD at p = 0.05 according to Duncan's multiple range test.

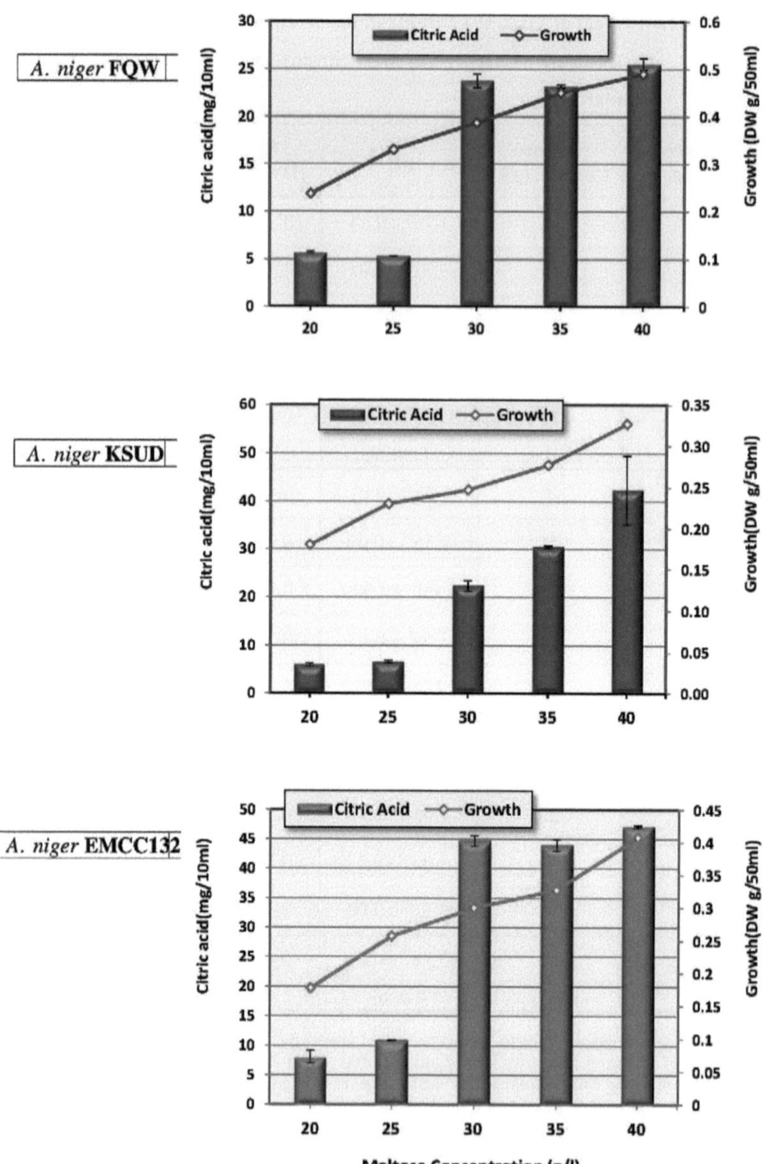

Fig. (4.III.6): Effect of maltose concentrations on citric acid production in shake flask culture of *Aspergillus niger* Strains.

Maddox *et al.,* (1985) reported the influence of different sources of carbon on citric acid production by *A. niger* and *Saccharomycopsis lipolytica.* Glucose, maltose, galactose, xylose and arabinose were tested. Fermentation was carried out in 8 and 4 days, respectively, at 30°C and 180 rpm. Better results were found for *A. niger* with glucose 27 g/L.

As well, **Auta *et al.,* (2014)** found that the concentration of 2% of the substrate could support citric acid production but an increase from 2% would result in higher levels of certain sugars present in the pulp that are inhibitory to citric acid production.

However, the present result is not comparable to the results reported by **Shu and Johnson (1948); Honecker *et al.,* (1989)** who found the final yield of citric acid increases when the initial sugar concentration is increased and maximal production rates achieved at 14–22% of sugar.

In addition, **Xu *et al.,* (1989)** studied the effect of carbohydrate concentration on citric acid yield in submerged fermentation. They tested maltose, sucrose, glucose, mannose, and fructose and they observed *Aspergillus niger* strains needed an initial sugar concentration of 10-14% as optimal, with the exception of glucose, where 7.5% gave the best results. No citric acid was produced on media containing less than 2.5% sugar. It was also noted that the increase in sugar concentration from 1% to 14%

increases the lag time for growth from 12 to 18 h and decreases the growth rate by around 20%.

Demirel *et al.,* **(2005)** reported maximum citric acid production at sucrose concentration of 140 g/L. As well, the optimum sugar concentration determined by **El-Hussein** *et al.,* **(2009)** was 150 g/L.

Similar conclusions were also reached by **Haq** *et al.,* **(2002) and Anwar** *et al.,* **(2009)** who observed significant reductions in citric acid yield when sugar concentration was increased beyond 150g/L. The latter authors explained this reduction by overgrowth of the mycelia resulting in increased viscosity of the medium.

From results of the present experiments, four percent of maltose concentration has been recommended for the maximal citric acid production from *A. niger* FQW, KSUD and EMCC132 strains and was selected as a control for the subsequent experiments.

4.3.7. Effect of Nitrogen Sources

Nitrogen constituent has a profound effect on citric acid production because nitrogen is not only important for metabolic rates in the cells but it is also basic part of cell proteins. This report agreed with **Grewal and Kalra (1995)** that fermentation media for citric acid biosynthesis should consist of substrates necessary for the growth of microorganism primarily the carbon, nitrogen and phosphorus sources.

This experiment was performed to evaluate the influence of different organic (urea, peptone, beef extract) and inorganic (sodium nitrate as a control, ammonium nitrate, ammonium chloride, ammonium sulfate) nitrogen sources on citric acid production by tested strains of *A. niger* (FQW, KSUD and EMCC132) at the previous favorable parameters.

The results were given in Table (4.III.7) and graphed in Figure (4.III.7) demonstrated that peptone as an organic nitrogen significantly enhanced citric acid production and showed superiority over all organic and inorganic nitrogen sources. Where maximum citric acid yield were reached to 49.78, 56.07 and 52.23 mg/10 ml by *A. niger* FQW, KSUD and EMCC132, respectively. Mostly presences of organic nitrogen sources acted as enhancer for both citric acid production and fungal growth.

However, inorganic nitrogen supplements had varied impact on production of the citric acid by all tested strains. Moreover, ammonium chlorid significantly suppressed citric acid production that gave only 5.53, 5.32 and 3.65 mg/10 ml by *A. niger* FQW, KSUD and EMCC132, respectively (as shown in Table 4.III.7 and Fig. 4.III.7).

Mostly complex nitrogen sources were better than inorganic nitrogen sources. This may be due to other nutrients and growth enhancers present in such organic compounds.

In addition, mycelia growth had been found high and reached to a maximum values in the presence of organic nitrogen source especially peptone and lowest fungal growth in the presence of inorganic nitrogen source, ammonium chloride.

Citric acid production is directly influenced by the nitrogen source. Physiologically, ammonium salts are preferred, e.g. urea, ammonium sulfate, peptone, malt extract, etc. Acid ammonium compounds are preferred because their consumption leads to pH decrease, which is essential for the citric fermentation. However, it is necessary to maintain pH values in the first day of fermentation prior to a certain quantity biomass production. Urea has a tampon effect, which assures pH control **(Kubicek and Rohr 1986)**.

Table (4.III.7): Citric acid production by *Aspergillus niger* strains on various nitrogen sources in shake flask culture. Data represented as mean ± standard error.

A. niger Strains	Nitrogen Sources	pH	Growth (DW g/50 ml)	Citric Acid (mg/10ml)
FQW	Control (Sodium nitrate)	2.96^b ±0.08	0.68^b ±0.02	25.63^b ±0.53
	Ammonium nitrate	2.69^c ±0.02	0.58^c ±0.02	11.03^d ±0.69
	Ammonium chloride	1.99^d ±0.00	0.28^g ±0.00	5.53^f ±0.01
	Ammonium sulfate	2.10^b ±0.01	0.47^e ±0.01	10.41^{de} ±0.36
	Beef extract	2.88^{bc} ±0.06	0.41^f ±0.02	9.39^e ±0.06
	Peptone	2.77^{bc} ±0.02	0.85^a ±0.02	49.78^a ±0.53
	Urea	4.25^a ±0.16	0.53^d ±0.01	18.12^c ±0.91
	LSD $_{0.05}$	0.2157	0.0463	1.6328
KSUD	Control (Sodium nitrate)	2.51^b ±0.04	0.62^b ±0.00	29.68^d ±0.25
	Ammonium nitrate	2.77^{cd} ±0.01	0.41^c ±0.04	12.33^e ±0.04
	Ammonium chloride	2.09^e ±0.02	0.22^e ±0.01	5.32^f ±0.03
	Ammonium sulfate	2.86^c ±0.02	0.28^{de} ±0.00	6.27^f ±0.12
	Beef extract	2.59^{bd} ±0.06	0.64^b ±0.01	50.05^b ±0.49
	Peptone	2.62^{bd} ±0.02	0.82^a ±0.03	56.07^a ±0.18
	Urea	3.38^a ±0.14	0.29^d ±0.02	43.27^c ±0.89
	LSD $_{0.05}$	0.1870	0.0588	1.2251
EMCC132	Control (Sodium nitrate)	3.84^b ±0.07	0.52^d ±0.01	47.00^b ±0.16
	Ammonium nitrate	3.10^d ±0.09	0.46^d ±0.01	13.97^d ±0.65
	Ammonium chloride	2.06^f ±0.02	0.29^f ±0.00	3.65^e ±0.30
	Ammonium sulfate	2.84^e ±0.03	0.68^b ±0.01	48.59^b ±0.71
	Beef extract	3.17^d ±0.02	0.37^e ±0.00	12.55^d ±0.67
	Peptone	3.33^c ±0.08	0.74^a ±0.00	52.23^a ±2.58
	Urea	4.07^a ±0.01	0.57^c ±0.01	25.86^c ±0.04
	LSD $_{0.05}$	0.1608	0.01793	3.2708

In modified Czapek-Dox broth medium. Initial pH was 6.5. Inoculum density was ~2.0X10^6 spore/ml. Cultures were grown with shaking at 100 rpm at 28°C for 5 days.
Means in the same column followed by the same letter are not significantly different based on LSD at p = 0.05 according to Duncan's multiple range test.

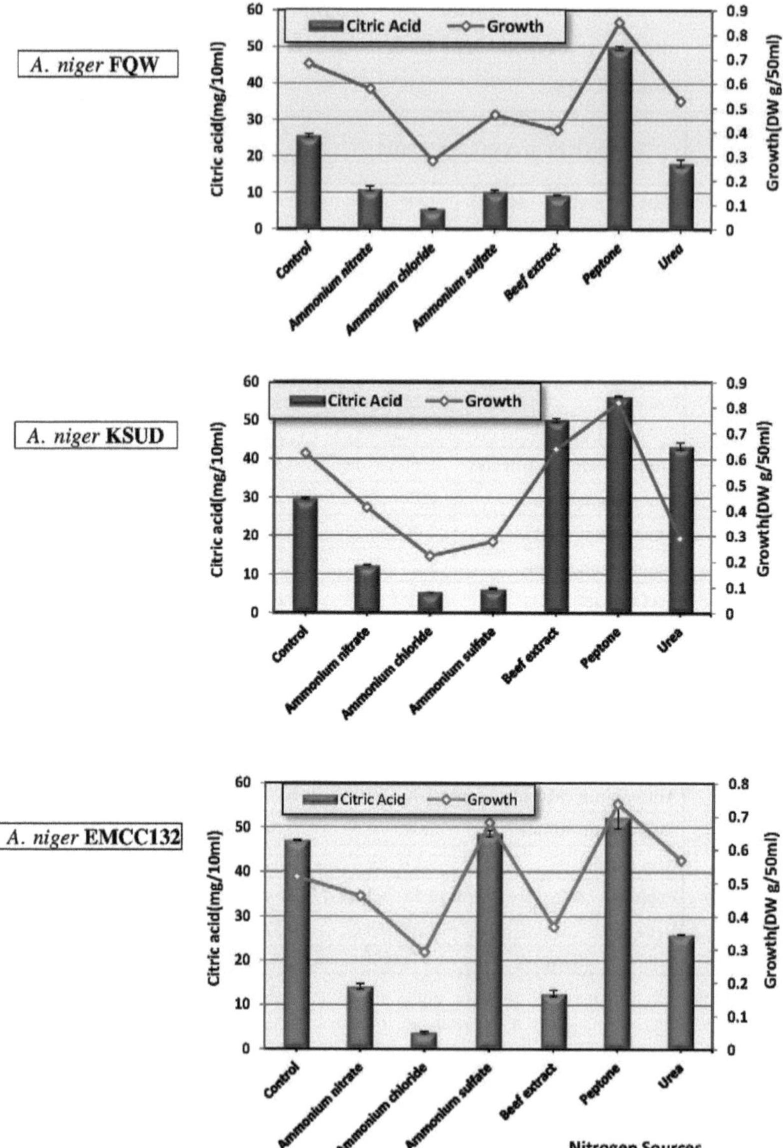

Fig. (4.III.7): Citric acid production by *Aspergillus niger* strains on various nitrogen sources in shake flask culture.

The effect of nitrogen source on citric acid production has been intensively studied in solid substrate and submerged fermentation. Ammonium chloride, ammonium sulphate, ammonium nitrate, pepton and yeast extract were the most suitable nitrogen source for production of citric acid by fungus (**Abou-Zeid and Ashy 1984**). Other sources of nitrogen that have been used include urea and yeast/malt extract (**Xu *et al.,* 1989**).

Complex media such as molasses are rich in nitrogen-containing compounds and rarely need to be supplemented with a nitrogen source. The high purity media that are used mainly in research laboratories are generally supplemented with ammonium salts, particularly ammonium nitrate and ammonium sulfate, to provide the necessary nitrogen. An advantage of using ammonium salts is that the pH declines as the salts are consumed and a low pH is a requirement of citric acid fermentation (**Mattey 1992**).

4.3.8. Effect of Nitrogen Source Concentrations

Nitrogen and its concentration in the growing media have been reported to be an important factor in fermentation processes due to an increase in C/N ratio. Thus, it is important to study the effectiveness of various concentrations of the nitrogen element on the behavior of the tested fungal strains and citric acid production. In that case, all of the previous studied parameters were utilized as optimum conditions that led to enhance of citric acid production by *A. niger* local strains FQW and *A. niger* KSUD up to 49.78 and 56.07 mg/10 ml, respectively (Table 4.III.7).

Impact of different concentrations of favorite nitrogen source, peptone (1.0, 2.0, 3.0, 4.0 and 5.0 g/l) on the production of citric acid production by all tested *A. niger'* strains (*A. niger* FQW, KSUD and EMCC132) was tested in flask cultures that adjusted to pH 6.5 and incubated in shaker incubator (100 rpm) at 28°C for 5 days.

The results represented in Table (4.III.8) and Figure (4.III.8) indicates that the citric acid yields proportional produced by concentrations increment, and the citric acid production was sharply jumped to the maximum values 50.52 mg/10 ml by *A. niger* KSUD at 3 g/l maltose concentration.

Table (4.III.8): Effect of peptone concentrations on citric acid production in shake flask culture of *Aspergillus niger* strains. Data represented as mean ± standard error.

A. *niger* Strains	Peptone Concentration (g/l)	pH	Growth (DW g/50 ml)	Citric Acid (mg/10ml)
FQW	1.0	2.80^a ±0.04	0.30^d ±0.04	10.43^c ±0.05
	2.0	2.79^a ±0.01	0.35^{cd} ±0.01	11.14^c ±0.01
	3.0	2.66^{bc} ±0.04	0.41^{bc} ±0.01	50.87^{ab} ±0.49
	4.0	2.74^{ab} ±0.04	0.45^b ±0.00	50.95^{ab} ±0.11
	5.0	2.62^c ±0.00	0.59^a ±0.05	51.67^a±0.25
	LSD $_{0.05}$	0.3432	0.0506	1.3881
KSUD	1.0	2.76^a ±0.02	0.39^c ±0.03	6.53^d ±0.36
	2.0	2.56^b ±0.00	0.50^b ±0.02	8.72^c ±0.15
	3.0	2.43^d ±0.01	0.52^b ±0.00	50.52^a ±0.22
	4.0	2.49^c ±0.01	0.64^a ±0.02	51.17^a ±0.51
	5.0	2.36^e ±0.02	0.68^a ±0.04	48.44^b ±0.14
	LSD $_{0.05}$	0.03669	0.07367	0.9728
EMCC132	1.0	3.48^a ±0.01	0.25^d ±0.04	6.67^d ±0.01
	2.0	3.43^{ab} ±0.04	0.34^c ±0.00	7.21^d ±0.08
	3.0	3.28^{bc} ±0.01	0.48^b ±0.00	50.89^c ±0.31
	4.0	3.50^a ±0.00	0.51^b ±0.02	52.56^b ±0.25
	5.0	3.23^c ±0.14	0.61^a ±0.02	61.03^a ±0.42
	LSD $_{0.05}$	0.1993	0.05828	0.8230

In modified Czapek-Dox broth medium. Initial pH was 6.5. Inoculum density was ~2.0X10^6 spore/ml. Cultures were grown with shaking at 100 rpm at 28°C for 5 days.
Means in the same column followed by the same letter are not significantly different based on LSD at p = 0.05 according to Duncan's multiple range test.

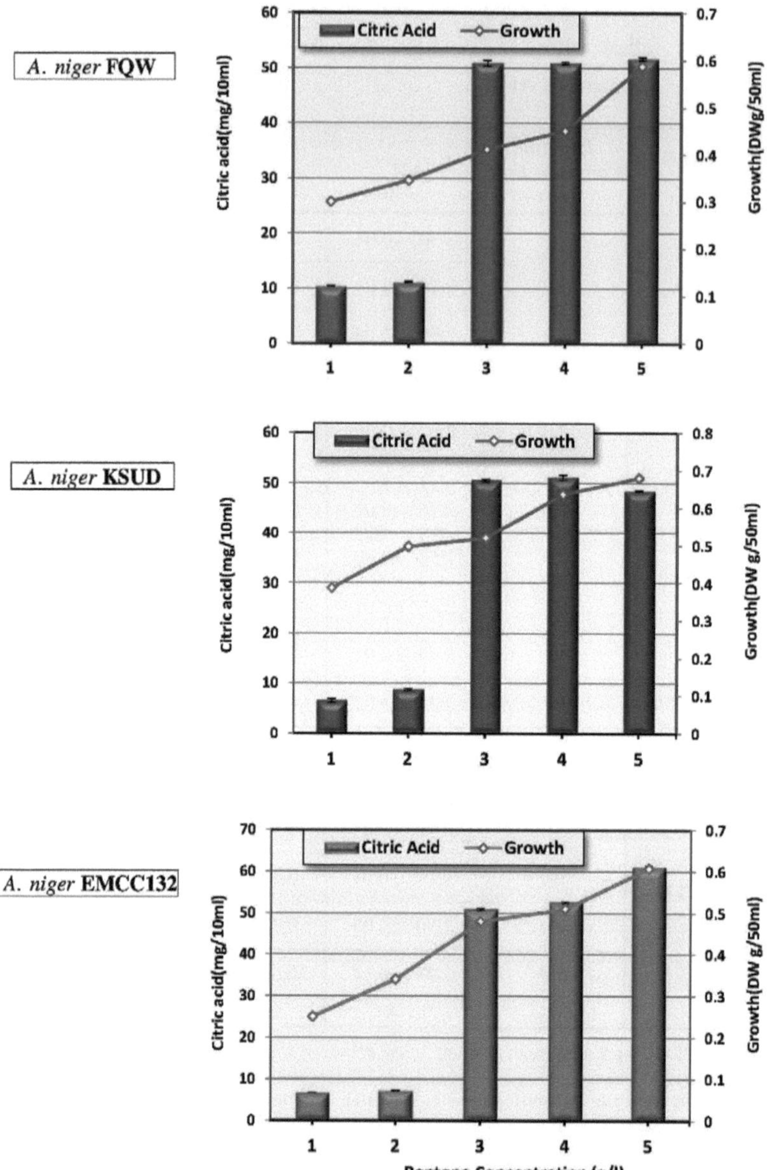

Fig. (4.III.8): Effect of peptone concentrations on citric acid production in shake flask culture of *Aspergillus niger* strains.

However in the case of *A. niger* FQW, citric acid yield reached to 50.17 mg/10 ml at 3.0 g nitrogen per liter and over this concentration (4.0 g/l) citric acid production not significantly increased, even at 5.0 g nitrogen per liter a negligible amount of citric acid produced.

In addition, it was found that mycelial dry weight of all tested *A. niger* strains were significantly enhanced with increasing of initial nitrogen concentration.

From the obtained results concerning the favorable maltose peptone concentration 3.0 g/l for maximal citric acid production, are in a range with certain previous researches.

The concentration of nitrogen source required for citric acid fermentation is 4 g N /liter. High nitrogen concentration increases fungal growth and the consumption of sugars, but decreases the amount of citric acid produced (**Hang *et al.*, 1977, Yokoya 1992; Vandenberghe *et al.*, 1999**).

Moreover, the limitation or starvation of nitrogen during the fermentation resulted in the limited growth of *A. niger* and to the enhancement of citric acid production (**Mirminachi *et al.*, 2002**).

4.3.9. Effect of Phosphorus Sources

Presence of phosphate in the medium has a great effect on the yield of citric acid. In the previous works it was verified that the presence of phosphate in the medium had a great effect on the yield of citric acid. Low levels of phosphate have positive effect on citric acid production. This effect acts at the level of enzyme activity and not at the level of gene expression. On the other hand, the presence of excess of phosphate leads to a decrease in the fixation of carbon dioxide, which in turn increases the formation of certain sugar acids, and the stimulation of growth (**Kubicek and Rohr 1986; Grewal and Kalra 1995; Vandenberghe *et al.,* 1999**).

Several factors affect citric acid production and among of these, phosphorus source has been found to play an important role in citric acid biosynthesis. So for enhancing the production of citric acid in the present study, *A. niger'* strains (*A. niger* FQW, KSUD and EMCC132) were cultivated on the Czapek-Dox broth medium amended with different phosphorus sources (NaH_2PO_4, Na_2HPO_4, Na_3PO_4 and KH_2PO_4) in the same amount of the initial phosphorus source, K_2HPO_4 (1.0 g/l) in flask shack culture at 28°C for 5 days to search for inducers of the citric acid overproduction.

Data represented in Table (4.III.9) and Figure (4.III.9) indicated that Na_3PO_4 significantly employed as the preferred phosphorus source for citric acid production by *A. niger* FQW (96.60 mg/10 ml) and *A. niger* KSUD (77.81 mg/10 ml) however in the case of reference strain *A. niger* EMCC132, KH_2PO_4 was significantly as the preferred carbon sources for citric acid production, 91.33 mg/10 ml.

All of the other forms of phosphate showed lower citric acid yields. Moreover, phosphate in the form of Na_2HPO_4 had shown significantly an inhibitory effect on citric acid production. In general, Na_3PO_4 served as the best phosphate form for citric acid production by local tested *A. niger'* strains (*A. niger* FQW and KSUD).

It is clear that citric acid production by fungal tested strains is dependent on nutritional conditions including phosphorus salt source. The fungal biomass was varied widely among the various phosphorus sources utilized (Table 4.III.9 and Fig. 4.III.9).

From results of the present experiment, phosphorus salt in the form Na_3PO_4 has been recommended for the maximal citric acid production from *A. niger* FQW, KSUD and EMCC132 strains and was selected as a control for the subsequent experiments.

Table (4.III.9): Citric acid production by *Aspergillus niger* strains on various phosphorus sources in shake flask culture. Data represented as mean ± standard error.

A. *niger* Strains	Phosphorus Sources	pH	Growth (DW g/50 ml)	Citric Acid (mg/10ml)
FQW	Control (K_2HPO_4)	2.54^a ±0.04	0.76^a ±0.00	55.60^d ±0.98
	Na_2HPO_4	2.43^b ±0.03	0.84^a ±0.06	16.66^e ±0.09
	Na_3PO_4	2.44^b ±0.02	0.70^a ±0.04	96.60^a ±0.49
	NaH_2PO_4	2.46^b ±0.01	0.78^a ±0.12	75.53^c ±0.90
	KH_2PO_4	2.45^b ±0.01	0.71^a ±0.01	80.91^b ±0.34
	LSD $_{0.05}$	0.0758	0.2351	2.0615
KSUD	Control (K_2HPO_4)	2.33^a ±0.01	0.74^c ±0.01	52.26^c ±0.35
	Na_2HPO_4	2.24^c ±0.00	0.86^b ±0.01	10.93^e ±0.12
	Na_3PO_4	2.27^{bc} ±0.03	1.01^a ±0.01	77.81^a ±1.79
	NaH_2PO_4	2.25^{bc} ±0.00	0.86^b ±0.06	68.41^b ±0.42
	KH_2PO_4	2.29^{ab} ±0.02	0.89^b ±0.01	47.86^d ±0.34
	LSD $_{0.05}$	0.0440	0.0941	2.6519
EMCC132	Control (K_2HPO_4)	2.86^{bc} ±0.015	0.54^c ±0.03	53.30^c ±0.76
	Na_2HPO_4	2.90^b ±0.02	0.55^{bc} ±0.04	11.60^d ±0.08
	Na_3PO_4	2.86^{bd} ±0.01	0.72^a ±0.00	81.31^b ±0.39
	NaH_2PO_4	2.82^{cd} ±0.02	0.62^b ±0.03	81.46^b ±0.47
	KH_2PO_4	2.96^a ±0.00	0.02^d ±0.01	91.33^a ±0.44
	LSD $_{0.05}$	0.0488	0.0799	1.5038

In modified Czapek-Dox broth medium. Initial pH was 6.5. Inoculum density was ~$2.0X10^6$ spore/ml. Cultures were grown with shaking at 100 rpm at 28°C for 5 days.
Means in the same column followed by the same letter are not significantly different based on LSD at p = 0.05 according to Duncan's multiple range test.

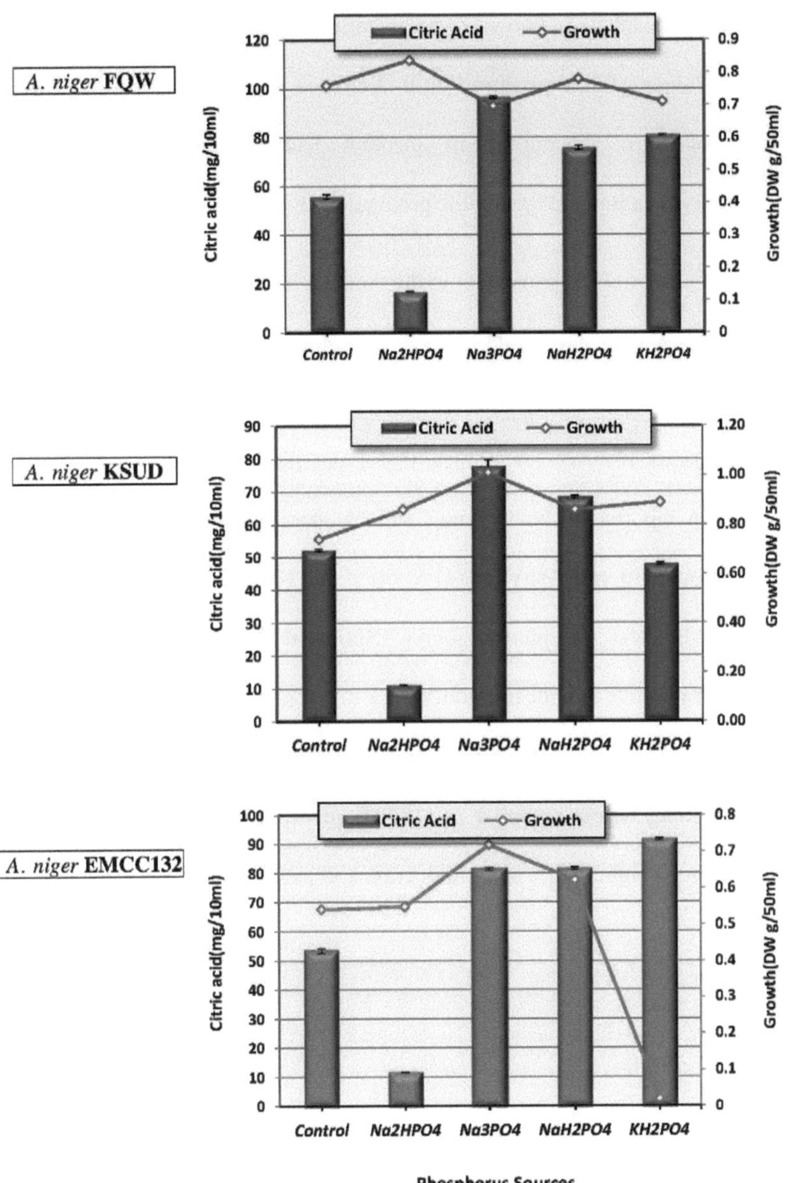

Fig. (4.III.9): Citric acid production by *Aspergillus niger* strains on various phosphorus sources in shake flask culture.

Phosphate is known to be essential for the growth and metabolism of *A. niger* (**Shankaranand and Lonsane 1994**). The concentration of exogenous phosphorus in medium had significant effect on cell multiplication and metabolite production (**Kubicek and Rohr 1986**).

Presence of phosphate in the medium has a great effect on the yield of citric acid. In the present study, phosphorus salt in the form Na_3PO_4 was the favorite phosphorus source by the tested fungal stains however certain previous research mentioned that Potassium dihydrogen phosphate has been reported to be the most suitable phosphorous source. For instance **Abou-Zeid and Ashy (1984)** reported that KH_2PO_4 and K_2HPO_4 proved to be the best phosphorus source. **Shu and Johnson (1948)** found that phosphorous at concentration of 0.5 to 5.0 g/L was required by the fungus in a chemically defined medium for maximum production of citric acid. In addition, **Mirminachi *et al.*, (2002)** reported that phosphorus limitation induced higher citric acid production and yield.

4.3.10. Effect of Enhancer (Alcohol Addition)

Among all of the previous studied parameters were utilized as optimum conditions which led to enhance of citric acid production by *A. niger* local strains FQW and *A. niger* KSUD (Table 4.III.9).

Besides the optimization of basal environmental parameters and nutrients as well as their concentrations, higher citric acid production levels can be obtained by applying stimulators, such as alcohols and chelating agents according to **Navaratnam *et al., (*1998)**. Several reports have shown the stimulatory effects of additives on fungal citric acid accumulation and secretion (**Kim 2004**). To improve citric acid production, stimulators have been used, such as organic solvents, phytate and lipids (**Jianlong and Ping, 1998; Kim 2004; Nadeem *et al.,* 2010**).

The objective of the present research was to determine the potential increase in citric acid production by *A. niger* selected strains (*A. niger* FQW, KSUD and EMCC132) in the presence of stimulators (1%) such as propanol, ethanol and methanol.

Data recorded in Table (4.III.10) and illustrated in Figure (4.III.10) indicate that after 5 days of fermentation, impact of alcohols addition on citric acid production by all tested *A. niger* strains was significantly varied.

Where methanol had obviously pronounced stimulatory effect, on contrary propanol was suppressed citric acid production while the effect of ethanol addition varied among the tested strains. Without methanol, the control produced 82.24 and 80.58 mg/10 ml while the addition of methanol significantly stimulated citric acid production to 128.2 and 103.58 mg/10 ml by *A. niger* FQW and KSUD, respectively.

By comparing the effect of both ethanol and methanol on biomass growth of tested local fungal strains it was observed that methanol has fruitful effect on biomass growth but there is no stimulatory effect of ethanol on biomass, which means that ethanol was solely used by *A. niger* FQW and KSUD as a carbon source for citric acid production. Similar findings were also reported by **Barrington and Kim (2008)**.

Alcohols have been shown to act principally on membrane permeability in microorganisms by affecting phospholipid composition. Other studies showed that alcohols stimulate citric acid production by affecting growth and sporulation on space organization of the membrane or changes in lipid composition of the cell wall **(Ingram and Buttke 1984)**.

The obtained results are in harmony with pervious data that showed the inductive effect of methanol in the fermentation media for citric acid by *A. niger* **(Hang and Woodams, 1998; Hang and Woodams, 1998 and Sukesh *et al.*, 2013)**.

Table (4.III.10): Citric acid production by *Aspergillus niger* strains in the presence of various alcohol sources (1%) in shake flask culture. Data represented as mean ± standard error.

A. niger Strains	Alcohol Sources	pH	Growth (DW g/50 ml)	Citric Acid (mg/10ml)
FQW	Control	2.61^b ±0.04	0.50^b ±0.00	82.24^c ±0.80
	Propanol	3.01^a ±0.07	0.29^c ±0.02	12.66^d ±0.13
	Ethanol	2.72^b ±0.03	0.45^b ±0.03	101.16^b ±0.99
	Methanol	2.68^b ±0.01	0.63^a ±0.06	128.20^a ±0.75
	LSD $_{0.05}$	0.1372	0.1080	2.4110
KSUD	Control	2.43^d ±0.02	0.56^b ±0.01	80.58^b ±0.90
	Propanol	3.21^a ±0.02	0.26^d ±0.00	8.61^d ±0.22
	Ethanol	2.54^c ±0.00	0.42^c ±0.00	73.20^c ±0.81
	Methanol	2.67^b ±0.01	0.68^a ±0.03	103.58^a ±3.74
	LSD $_{0.05}$	0.0456	0.0515	6.415
EMCC132	Control	2.90^c ±0.01	0.75^b ±0.02	79.89^b ±1.03
	Propanol	3.42^a ±0.02	0.32^d ±0.01	14.79^c ±0.07
	Ethanol	3.42^a ±0.03	0.44^c ±0.02	15.35^c ±0.09
	Methanol	3.02^b ±0.00	0.92^a ±0.01	90.30^a ±0.48
	LSD $_{0.05}$	0.05324	0.0492	1.8600

In modified Czapek-Dox broth medium. Initial pH was 6.5. Inoculum density was ~2.0X10^6 spore/ml. Cultures were grown with shaking at 100 rpm at 28°C for 5 days.
Means in the same column followed by the same letter are not significantly different based on LSD at p = 0.05 according to Duncan's multiple range test.

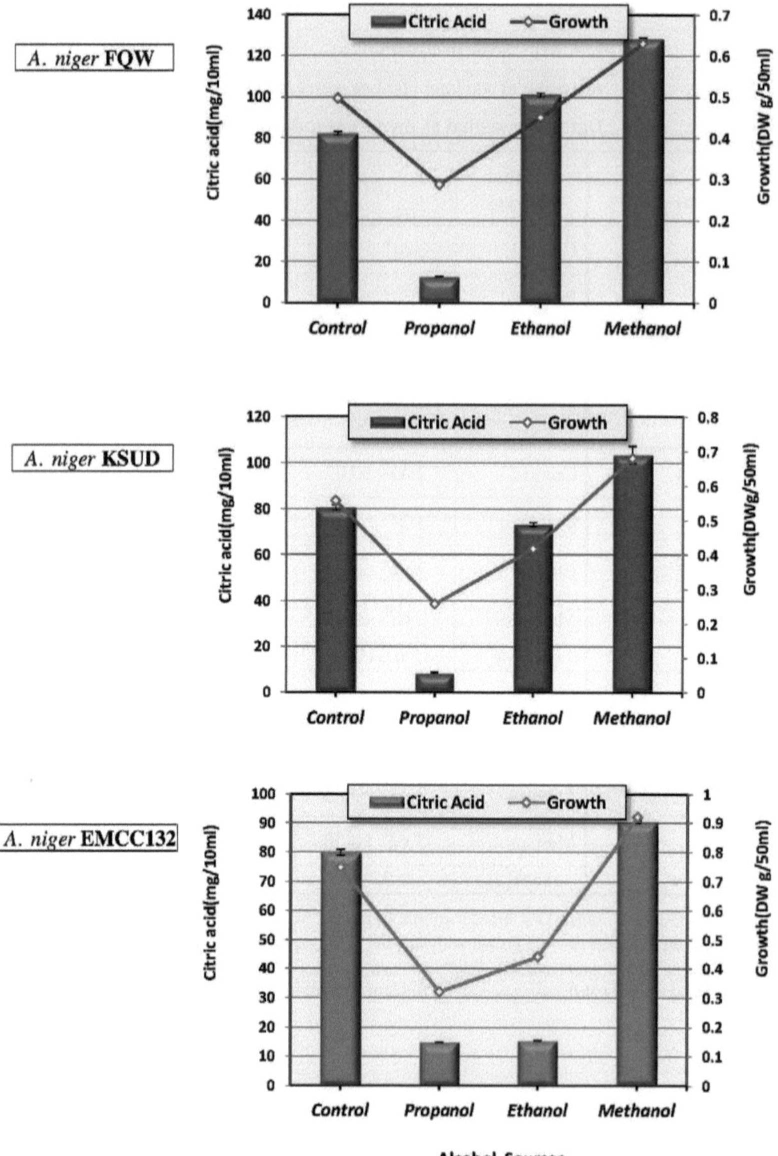

Fig. (4.III.10): Citric acid production by *Aspergillus niger* strains in the presence of various alcohol sources (1%) in shake flask culture.

Where several hypotheses related to the stimulating effect of methanol on citric acid production were discussed, the increase in citric acid production may result from a decrease in mycelial growth; glucose was used for citric acid production rather than for cell growth (**Hang and Woodams 1998**). Also, organic solvents such as alcohol e.g. methanol are known to increase the permeability of the cell membrane by affecting phospholipid composition on the cytoplasmatic membrane and, thus, improve the secretion of citric acid across this membrane (**Orthofer *et al.*, 1979; Navaratnam *et al.*, 1998**). However **Meixner *et al.*, (1985)** argued against a role of membrane permeability in citric acid accumulation. Also **Ingram and Buttke (1984)** found that alcohols stimulate citric acid production by affecting growth and sporulation through the action not only on the cell permeability but also the spatial organization of the membrane, or changes in lipid composition of the cell wall.

Nadeem *et al.*, (2010) mentioned that the stimulation of citric acid production by methanol in synthetic media is affected by cultural conditions, and especially by the mold strain used. The age and the amount of mycelia inoculum which is probably a reflection of its surface area may be critical. They reported that these factors must be investigated in applying the effect of the alcohol in any individual case.

Referring to the obtained results of ethanol effectiveness on citric acid production, the effect of ethanol addition varied among the tested strains. However, these finding is in contrary with pervious data that showed the enhancement effect of ethanol in the fermentation media for citric acid.

Several researchers have reported the stimulating effect of ethanol on citric acid production and suggested that ethanol could be used as a carbon source to be converted to citric acid via the TCA cycle of *A. niger* (**Navarantnam** *et al.,* **1998**). In addition to its assimilation, ethanol can also increase the permeability of the cell membrane and, thus, increase the secretion of citric acid (**Pazouki** *et al.,* **2000**).

From results of this experiment, methanol had significant favorable effect for the maximal citric acid production by local fungal strains (*A. niger* FQW and KSUD) and selected for the subsequent experiments.

4.3.11. Effect of Alcohol Concentrations

According to our study، methanol (1.0%) showed stimulatory effect on citric acid production by tested fungal strains (*A. niger* FQW, KSUD and EMCC132).

In order to improve the production of citric acid, *Aspergillus niger*' isolates were cultivated on modified Czapek-Dox broth medium which before inoculation supplemented with different methanol concentrations (1.0, 2.0, 3.0, 4.0 and 5.0%) in flask shack culture and at 28°C for 5 days to search for optimal concentration of inducer (methanol) of the citric acid biosynthesis.

Data represented in Table (4.III.11) and graphed in Fig. (4.III.11) indicate that by increasing alcohol concentrations, 2% of methanol were significantly enhanced citric acid biosynthesis and reached to the maximum yield 133.31, 109.29 and 149.74 mg/10ml by *A. niger* FQW, KSUD and EMCC132, respectively. Methanol concentration over 2% sharply deteriorated citric acid production by all tested fungal strains.

Fungal biomass of all tested fungal strains was varied widely among the methanol concentration utilized where mycelial yields were gradually raised and significantly gave highest values at 3% of methanol.

Table (4.III.11): Effect of methanol concentrations on citric acid production in shake flask culture of *Aspergillus niger* strains. Data represented as mean ± standard error.

A. *niger* Strains	Methanol Concentration (%)	pH	Growth (DW g/50 ml)	Citric Acid (mg/10ml)
FQW	0.0	3.15^a ±0.00	0.57^{bc} ±0.00	61.08^c ±0.36
	1.0	3.01^{bc} ±0.02	0.67^b ±0.10	129.01^b ±0.21
	2.0	3.16^a ±0.02	0.63^b ±0.09	133.31^a ±0.08
	3.0	3.09^{ab} ±0.03	0.89^a ±0.02	50.63^d ±0.64
	4.0	2.90^d ±0.06	0.44^c ±0.02	11.95^e ±0.07
	5.0	2.99^c ±0.01	0.28^d ±0.04	8.92^f ±0.05
	LSD $_{0.05}$	0.08195	0.1390	0.9806
KSUD	0.0	2.62^b ±0.04	0.51^b ±0.01	47.12^c ±0.53
	1.0	2.41^e ±0.01	0.39^c ±0.03	96.73^b ±2.71
	2.0	2.58^{bc} ±0.01	0.54^b ±0.00	109.29^a ±1.18
	3.0	2.54^{cd} ±0.00	0.85^a ±0.06	17.99^d ±0.11
	4.0	2.52^d ±0.01	0.23^d ±0.02	16.37^d ±0.35
	5.0	2.7^a±0.01	0.14^e ±0.01	15.15^d ±0.10
	LSD $_{0.05}$	0.0578	0.0879	3.8065
EMCC132	0.0	2.49^d ±0.01	0.40^{bc} ±0.00	84.79^b ±2.53
	1.0	2.67^c ±0.02	0.47^b ±0.01	137.42^a ±9.56
	2.0	2.68^c ±0.01	0.60^a ±0.04	149.74^a ±1.18
	3.0	2.68^c ±0.01	0.60^a ±0.04	149.74^a ±1.18
	4.0	3.13^b ±0.01	0.41^b ±0.01	62.79^c ±0.51
	5.0	3.44^a ±0.02	0.34^c ±0.00	53.27^c ±0.13
	LSD $_{0.05}$	0.0404	0.0739	12.5789

In modified Czapek-Dox broth medium. Initial pH was 6.5. Inoculum density was ~2.0×10^6 spore/ml. Cultures were grown with shaking at 100 rpm at 28°C for 5 days.
Means in the same column followed by the same letter are not significantly different based on LSD at p = 0.05 according to Duncan's multiple range test.

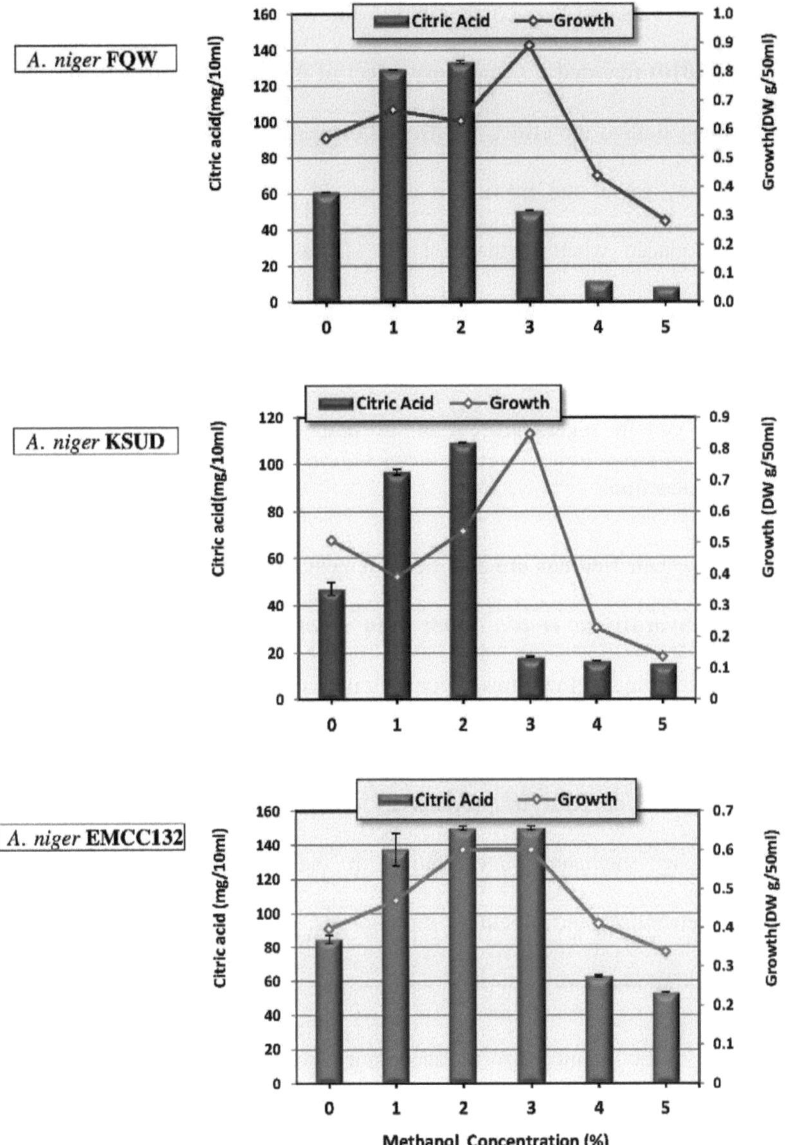

Fig. (4.III.11): Effect of methanol concentrations on citric acid production in shake flask culture of *Aspergillus niger* strains.

This result is comparable to the results reported by **Nadeem *et al.,* (2010)** reported a stimulatory effect of methanol on citric acid production and studied the effect of different concentrations of methanol (0.5-2.0 %). They found the maximum amount of citric acid (49.33±4.23 g/l) was produced when methanol (1.5 %) was added into the beet molasses medium. In addition the concentration of 2% of methanol could support citric acid production but an increase from 2% would result in higher levels of certain sugars present in the pulp that are inhibitory to citric acid production.

Our findings are also in agreement to those reported by investigators **(Navaratnam *et al.,* 1998; Haq *et al.,* 2003; Nadeem *et al.,* 2010)**. According to them lower concentration of methanol has stimulatory effect on citric acid production. However increase in concentration of methanol tends to decrease its productivity. This might be due to the fact that the higher methanol concentration in the medium disturbed the fungal metabolism and inoculum morphology, which resulted in decrease citric acid production.

The results are in accordance with other previous studies. For instance, **El-Hussein *et al.,* (2009)** showed that the effect of adding different concentrations of methanol on mycelial dry mass and citric acid production by *Aspergillus niger* isolate SUD27 were varied. Methanol had significant favorable effects on citric acid yield. At 3 % methanol, citric acid production was 76.81 g/L which is significantly better, by 37%, than the control.

Methanol neutralizes the negative effect of the metals in citric acid production generally in amounts about 1 to 5 % (**Kubicek and Rohr 1986**). Even so, optimal amount of methanol depends upon the strain and the composition of the medium.

Among all of the above results, it could be concluded the following fermentation parameters as summarized in Table (4.III.12) were selected as optimum for the production of citric acid using local fungal strains, (*Aspergillus niger* FQW and KSUD). The present study showed that local strains of *Aspergillus niger* have the potential to produce citric acid in large quantity at the optimized fermentation conditions.

Citric acid production was clearly improved through various studied parameters that gave 14-12 times higher and reached to 133.31 and 109.29 mg/10ml (Table 4.III.11) comparing to the original yields 9.07 and 9.98 mg/10ml (Table 4.II.1) by *Aspergillus niger* FQW and KSUD, respectively.

Table (4.III.12): Summary of the optimal values for maximal production of citric acid by local fungal strains, *Aspergillus niger* FQW and KSUD.

No.	Variable	Units/form	Optimum Value
1.	Incubation period	Days	5
2.	Fermentation Temperature	°C	28
3.	Initial pH	Value	6.5
4.	Inoculum density	Spores/ml	$3.0X10^6$
5.	Agitation	rpm	100
6.	Carbon source	Sugar	Maltose
7.	Carbon concentration	g/l	40
8.	Nitrogen source	Organic	Peptone
9.	Nitrogen concentration	g/l	3
10.	Phosphorus Source	Inorganic	Na_3PO_4
11.	Alcohol addition	Organic solvent	Methanol
12.	Alcohol concentration	%	2

4. 4. Fourth Part: Utilization of agricultural wastes for Citric Acid Production by *A. niger* strains

A. niger a filamentous fungus is most commonly used for citric acid production due to ease of handling, its ability to utilize varieties of substrates because of its well developed enzymatic system in addition to its ability to ferment a variety of cheep raw materials and high yield of citric acid (**Schuster *et al.*, 2002; Boominathan *et al.*, 2012; Femi-Ola and Atere 2013**).

The worldwide demand of citric acid is about 6.0×105 tons per year and it is bound to increase day by day. With an estimated annual production of 1,000,000 tons, citric acid is one of fermentation products with the highest production level worldwide. The food industry consumes about 70% of total citric acid produced, while other industries consume remaining 30% (**Milson 1987**). From this point of view, it is necessary to use inexpensive and readily available raw materials for industrial production processes.

Also, a reduction cost in citric acid production can be achieved by using cheap local agricultural wastes such as sugarcane bagasse; potato peel; pineapple peel; dates molasses and sugarcane molasses that were used in this study. In addition, the potential of agricultural wastes as a medium for the production of citric acid was investigated with a view to reducing its disposal problem and environmental pollution.

Moreover, utilization refined sugars such as glucose, maltose and sucrose are the most commonly used substrates for commercial production of citric acid by fermentation process; while they are expensive and can be replaced by various cheap and abundant substrates like agro-industrial wastes or by-products (**Prasad *et al.*, 2014**).

As presented previously, several raw materials can be employed successfully for citric acid production. However, there are some critical factors that should be taken into account such as costs or need of pretreatment for choosing the type of substrate (**Soccol *et al.*, 2006**).

Thus, the objective of this part was to adopt the use of certain local raw materials as a cheap medium for the production of citric acid by tested *A. niger*. In this production technique, which is still the major industrial rout to commercial citric acid production nowadays, tested strains of *A. niger* are cultivated on a raw material-containing modified Czapek-Dox broth medium to produce citric acid.

Citric acid productivity by *A. niger* tested strains (*A. niger* FQW, KSUD and EMCC132) was determined after 5 days growing cultures in which 5% (w/v or v/v) of selected raw materials (pineapple peel, sugarcane bagasse, sugarcane molasses and dates molasses) or 3% (w/v) of potato peel was supplemented to the fermentative medium either carbon free or carbon and nitrogen free.

The estimated results present in Table (4.IV.1) and illustrated in Figure (4.IV.1) indicate citric acid production by tested fungal strains in the presence of selected raw materials as a sole carbon source. The results showed that sugarcane molasses was significantly superior for citric acid production (256.94 and 188.45 mg/10 ml) by *A. niger* FQW and *A. niger* EMCC132, respectively. However, *A. niger* KSUD produced highest amount of citric acid (238.19 mg/10 ml) in the presence of raw material pineapple peel. It also obviously appeared that dry weight of mycelial growth for all tested strain significantly pronounced with the same raw material that gave highest citric acid production.

On contrary, utilization of potato peels showed raising in pH values (6.36, 7.10 and 6.73) and sharply decreased citric acid biosynthesis (11.07, 12.64 and 14.87 mg/10 ml) in corresponding with growth deterioration (0.22, 0.24 and 0.12 g/50ml) of all tested fungal strains *A. niger* FQW, KSUD and EMCC132, respectively.

In addition, in the presence of sugarcane molasses although all tested fungal strains mostly well grew and gave mycelial growth ranged in 0.46 – 0.62 g/50ml, citric acid significantly estimated lowest yields 6.61, 10.53 and 7.24 mg/10 ml by *A. niger* FQW, KSUD and EMCC132, respectively, moreover raising in pH values were measured (7.73, 6.20 and 7.63).

Refereeing to the pervious results of citric acid produced by tested fungal strains in the presence of used raw materials as a sole carbon sources in the growing cultures, it is indicated that citric acid yield was widely varied and ranged between 7.24-256.94 mg/10 ml. The used raw materials were enhanced citric acid production statistically in the following order: sugarcane molasses > pineapple peels > date molasses > potato peels > sugarcane bagasse by *A. niger* FQW and pineapple peels > sugarcane molasses > date molasses > potato peels ≥ sugarcane bagasse by *A. niger* KSUD as well sugarcane molasses > date molasses > pineapple peels > potato peels > sugarcane bagasse by reference strain *A. niger* EMCC132.

Local strain *A. niger* FQW produced highest amount of citric acid, reached up to 256.94 mg/10 ml when grew on sugarcane molasses as a sole carbon source.

By assessing the obtained findings among used raw materials, it is considering the fact that for production of commercially valuable citric acid by tested *A. niger* strains, molasses (either sugarcane or date) and pineapple peels proved to be appropriate substrates from the cost point of view, it is highly possible to consider these effective processes as beneficial, economical or cost-effective methods.

Table (4.IV.1): Citric acid production by *Aspergillus niger* strains grown in fermentative medium carbon free and supplemented with various waste sources. Data represented as mean ± standard error.

A. *niger* Strains	Waste Sources	pH	Growth (DW g/50 ml)	Citric Acid (mg/10ml)
FQW	Pineapple peels	3.83^c ±0.04	0.94^b ±0.06	180.25^b ±0.66
	Sugarcane bagasse	7.73^a ±0.13	0.46^d ±0.04	6.61^e ±0.51
	Sugarcane molasses	3.03^d ±0.11	1.74^a ±0.10	256.94^a ±1.09
	Date molasses	2.91^d ±0.05	0.71^c ±0.05	142.88^c ±0.38
	Potato peels	6.36^b ±0.19	0.22^e ±0.01	11.07^d ±0.16
	LSD $_{0.05}$	0.3723	0.1937	2.0161
KSUD	Pineapple peels	2.40^d ±0.04	1.33^a ±0.03	238.19^a ±1.14
	Sugarcane bagasse	6.20^b ±0.05	0.56^d ±0.02	10.53^d ±0.08
	Sugarcane molasses	2.56^c ±0.02	1.01^b ±0.03	168.87^b ±1.18
	Date molasses	2.50^{cd} ±0.00	0.71^c ±0.06	162.67^c ±0.34
	Potato peels	7.01^a ±0.01	0.24^e ±0.01	12.64^d ±0.11
	LSD $_{0.05}$	0.1016	0.1051	2.3762
EMCC132	Pineapple peels	3.36^c ±0.02	0.74^c ±0.02	127.73^c ±0.41
	Sugarcane bagasse	7.63^a ±0.12	0.62^d ±0.01	7.24^e ±0.11
	Sugarcane molasses	3.15^{cd} ±0.02	1.06^a ±0.01	188.45^a ±0.65
	Date molasses	2.91^d ±0.01	0.82^b ±0.03	182.14^b ±0.06
	Potato peels	6.73^b ±0.17	0.12^e ±0.01	14.873^d ±0.25
	LSD $_{0.05}$	0.29505	0.05478	1.1507

In modified Czapek-Dox broth medium. Initial pH was 6.5. Inoculum density was ~2.0X10^6 spore/ml. Cultures were grown with shaking at 100 rpm at 28°C for 5 days.
Means in the same column followed by the same letter are not significantly different based on LSD at p = 0.05 according to Duncan's multiple range test.

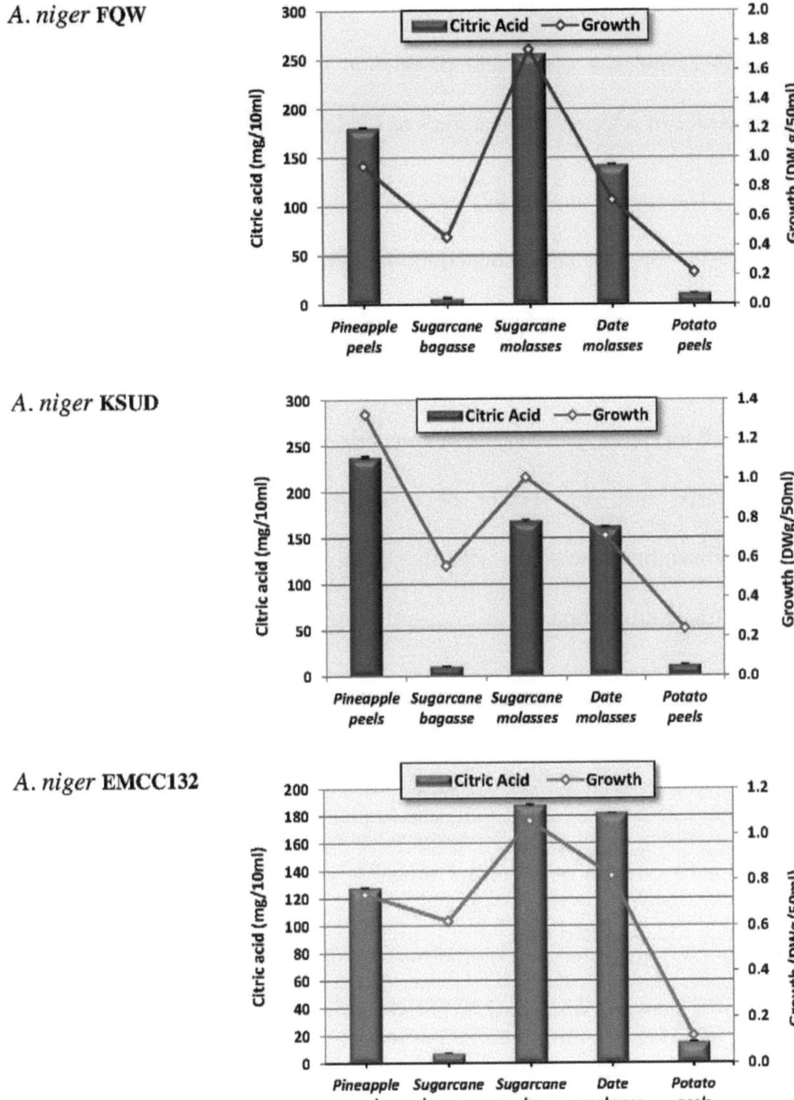

Fig. (4.IV.1): Citric acid production by *Aspergillus niger* strains grown in fermentative medium carbon free and supplemented with various waste sources.

The obtained results present in Table (4.IV.2) and illustrated in Figure (4.IV.2) indicate citric acid production by tested fungal strains in the presence of selected raw materials as a sole carbon and nitrogen source.

The results showed that raw material pineapple peel was significantly superior for citric acid production (133.23 and 223.77 mg/10 ml) by local strains *A. niger* FQW and *A. niger* KSUD, respectively. While, reference strain *A. niger* EMCC132 produced highest amount of citric acid (172.72 mg/10 ml) in the presence of raw material date molasses. It also obviously appeared that dry weight of mycelial growth for all tested strain significantly pronounced with the same raw material that gave highest citric acid production.

Potato peels utilization still showed raising in pH values (8.23, 8.12 and 8.33) which resulted in sharply decreasing citric acid biosynthesis (7.17, 7.25 and 9.32 mg/10 ml) and fungal growth deterioration (0.09, 0.07 and 0.10 g/50ml) of all tested fungal strains *A. niger* FQW, KSUD and EMCC132, respectively.

In general, when raw materials utilized in fermentation culture as carbon and nitrogen source, citric acid amounts obviously decreased comparing with the previous process when used it as sole carbon source, except sugarcane molasses utilization as sole carbon and nitrogen source led to stimulate citric acid production (Tables 4.IV.1 and 4.IV.2).

Table (4.IV.2): Citric acid production by *Aspergillus niger* strains grown in fermentative medium carbon & nitrogen free and supplemented with various waste sources. Data represented as mean ± standard error.

A. niger Strains	Waste Sources	pH	Growth (DW g/50 ml)	Citric Acid (mg/10ml)
FQW	Pineapple peels	$2.62^d \pm 0.00$	$0.72^a \pm 0.14$	$133.23^a \pm 0.16$
	Sugarcane bagasse	$2.58^d \pm 0.04$	$0.19^c \pm 0.02$	$85.12^d \pm 0.34$
	Sugarcane molasses	$4.34^b \pm 0.02$	$0.32^{bc} \pm 0.00$	$87.76^c \pm 1.12$
	Date molasses	$2.82^c \pm 0.00$	$0.40^b \pm 0.04$	$124.38^b \pm 0.40$
	Potato peels	$8.23^a \pm 0.04$	$0.09^c \pm 0.01$	$7.17^e \pm 0.03$
	LSD $_{0.05}$	0.0751	0.2046	1.7637
KSUD	Pineapple peels	$2.47^d \pm 0.00$	$0.88^a \pm 0.07$	$223.77^a \pm 1.43$
	Sugarcane bagasse	$2.63^c \pm 0.07$	$0.27^c \pm 0.02$	$92.67^d \pm 0.16$
	Sugarcane molasses	$3.01^b \pm 0.00$	$0.34^{bc} \pm 0.02$	$95.64^c \pm 0.13$
	Date molasses	$2.47^d \pm 0.00$	$0.42^b \pm 0.01$	$196.92^b \pm 0.44$
	Potato peels	$8.12^a \pm 0.06$	$0.07^d \pm 0.01$	$7.25^e \pm 0.03$
	LSD $_{0.05}$	0.1296	0.1016	2.1282
EMCC132	Pineapple peels	$2.67^d \pm 0.01$	$0.69^a \pm 0.00$	$116.02^b \pm 0.25$
	Sugarcane bagasse	$2.82^c \pm 0.01$	$0.19^d \pm 0.00$	$58.05^d \pm 0.28$
	Sugarcane molasses	$4.14^b \pm 0.06$	$0.28^c \pm 0.03$	$112.68^c \pm 1.12$
	Date molasses	$2.74^{cd} \pm 0.01$	$0.34^b \pm 0.01$	$172.70^a \pm 0.72$
	Potato peels	$8.34^a \pm 0.00$	$0.10^e \pm 0.01$	$9.32^e \pm 0.02$
	LSD $_{0.05}$	0.0855	0.0474	1.9461

In Czapek-Dox broth medium. Cultivation temperature was 28°C. Cultures were grown with shaking at 100 rpm. Cultivation period was 5 days.

Means in the same column followed by the same letter are not significantly different based on LSD at p = 0.05 according to Duncan's multiple range test.

A. *niger* **FQW**

A. *niger* **KSUD**

A. *niger* **EMCC132**

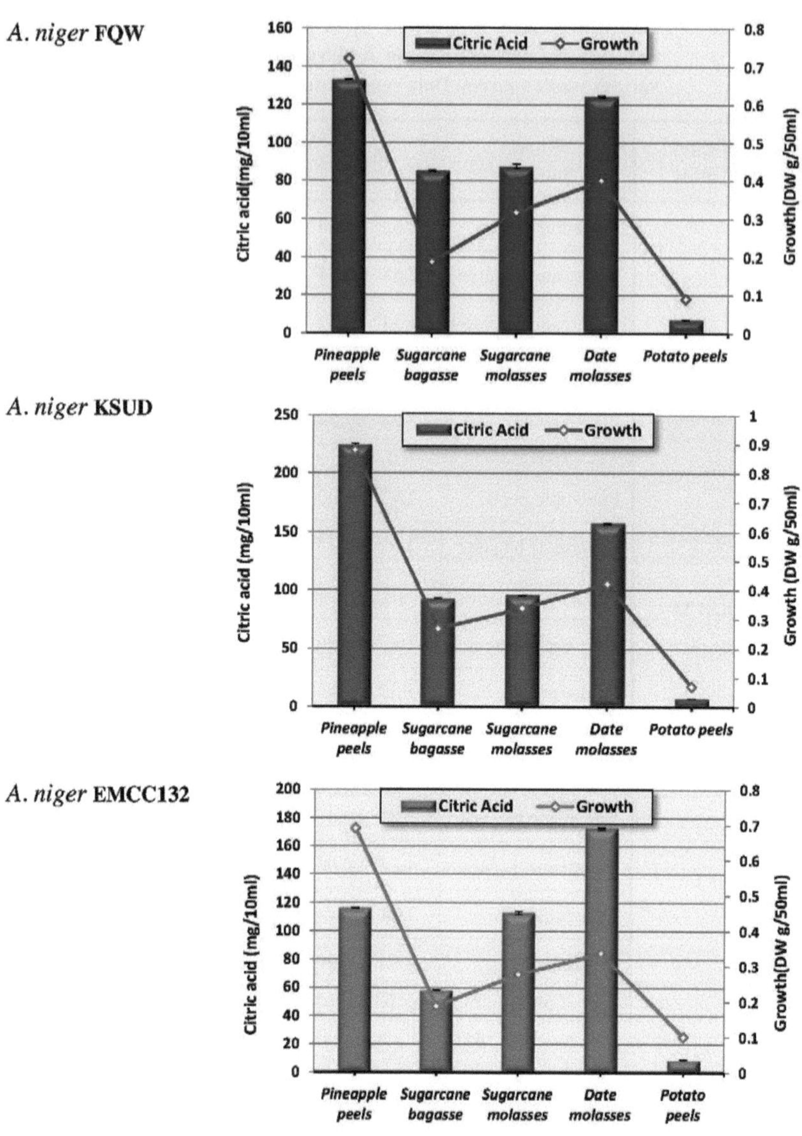

Fig. (4.IV.2): Citric acid production by *Aspergillus niger* strains grown in fermentative medium carbon & nitrogen free and supplemented with various waste sources.

In addition, local strain *A. niger* KSUD produced highest amount of citric acid, reached up to 223.77 mg/10 ml when grew on pineapple peels as a sole carbon and nitrogen source.

So it is highly recommended that the usage of sugarcane molasses waste supplemented medium with *A. niger* will be effective in the production of citric acid than the other substrates.

The maximum yield of citric acid after 5 days fermentation in the previous research were summarized in Table (4.IV.3) and compared with citric acid production in current study.

Several previous in harmony with these finding which reported for commercial production of citric acid is generally by submerged fermentation of molasses using the filamentous fungus *A. niger* or **(Chaudary *et al.*, 1978; Dasgupta *et al.*, 1994; Adham 2002; Ali *et al.*, 2002; Kumar *et al.*, 2003; Ali 2004; Haq *et al.*, 2004; El-Aasar 2006)**.

Complex media such as molasses are rich in nitrogen-containing compounds and rarely need to be supplemented with a nitrogen source. The high purity media that are used mainly in research laboratories are generally supplemented with ammonium salts to provide the necessary nitrogen **(Mattey, 1992)**.

Table (4.IV.3): Comparison of citric acid production between current research results and the previous literature values.   167

Fungal strain	Substrate	Fermentation process	Citric acid yield (g/L)	Reference
Aspergillus niger FQW	Sugarcane molasses	Shake flask	25.69	Present study
Aspergillus niger KSUD	pineapple peels	Shake flask	23.82	Present study
Aspergillus niger 14/20	Molasses	Shake flask	7.72	Majumder *et al.*, (2009)
Aspergillus niger	Bagasse	Shake flask	0.24	Vaishnavi *et al.*, (2012)
Aspergillus niger	Beet molasses	Shake flask	240.10	Lotfy *et al.*, (2007a)
Aspergillus niger ATTCC 9142	Beet molasses	Shake flask	35.00	Roukas (1991)

Molasses is preferably used as the source of sugar for microbial production of citric acid due to its relatively low cost and high sugar content 40–55 % (**Grewal and Kalra 1995**). Since it is a by-product of sugar refining, the quality of molasses varies considerably, and not all types are suitable for citric acid production. The molasses composition depends on various factors like the variety of beet and cane, methods of cultivation, conditions of storage and handling (transport, temperature variations), etc. Both beet and cane molasses are suitable for citric acid production, however, beet molasses is preferred to sugarcane due to its lower content of trace metals, supplying better production yields than cane molasses, but there are considerable yield variations within each type. In the case of cane molasses, generally it contains some metals (iron, calcium, magnesium, manganese, zinc) which may retard citric acid synthesis and it requires some pretreatment for the reduction of them (**Soccol *et al.*, 2006; El-Hussein *et al.*, 2009**).

Pineapple peel is a by-product resulting from the processing of pineapple into slices and represents about 10% w/w of the weight of the original fruit. Current disposal of it poses considerable economic and environmental problems (**Kareem *et al.*, 2010**). They found that pineapple waste with 15% (w/v) sucrose and ammonium nitrate (0.25% w/v) gave the optimum citric acid secretion (60.61 g/kg) in the presence of methanol (2%

v/v) when fermented for 5 days at 30 °C with the initial moisture content of 65%. Their results proved the use of pineapple peel as a cheap medium for the production of commercially valuable organic acid by *A. niger*.

Although, sugarcane bagasse was failed to produce citric by all tested fungal strains in submerged culturing, **Yadegary *et al.,* (2013)** indicated that sugarcane bagasse is an appropriate substrate in producing citric acid and the aforementioned process could be considered as a beneficial and cost-effective method in citric acid production. They also found the highest yields of citric acid production from sugarcane bagasse were obtained through the addition of urea up to 4% (v/v) as a nitrogen source.

Most of the raw materials used in current study, citric acid production were reduced in fermentative media free of nitrogen. It could be explained according to **Alam *et al.,* (2001)** who reported that when *A. niger* is added to a medium which is either low in or free of nitrogen, it produces as much oxalic acid as citric acid. However, an increase in nitrogen source produces opposite results and it causes an increase in microorganism's growth and a decrease in citric acid production.

Faruk *et al.* (2014) studied the effect of potato peels hydrolysate on the mycelial growth and citric acid production of *Aspergillus niger* in medium containing sucrose and Prescott salt in shake flasks. They obtained 6-fold increment in biomass and 5-fold increment in citric acid production after 13th day of fermentation. Supplementation dried potato peels reduced fermentation media cost.

At the present, the fundamental exploitation of food waste, which participates in pollution, is the controlled biological degradation of the wastes by microorganisms for the production of valuable compounds such as enzymes, citric acid and others as raw materials for medical and industrial uses.

Keeping in consideration when agricultural wastes were used, to be as cheaper and effective for production of citric acid. Because of the present study deals with production and optimization of citric acid by using economically feasible and easily available natural substrates.

4. 5. Fifth Part: Citric acid production by mutated *A. niger* strains

The objective of this part is an another perspective to improve the citric acid productivity by using muted *Aspergillus niger* strains compared with wild strains. We are supposing and hoping that mutation will be one of the useful methods for improvement *A. niger* wild strains and subsequently increase the production of citric acid.

Mutations are abrupt and so are the hereditary modifications in the genetic material. The organisms containing the DNA are not static molecules and their bases are frequently exposed to natural or artificial agents that can cause modifications in their structure or in chemical composition (**Zaha, 2003**).

Mutation was carried out in order to develop mutants of parent strain for increased production of the products (**Adham, 2002**). However, strain development from wild strains to mutants depends mainly on the process of mutagenesis (physical and chemical mutagenesis). Although, traditional random strain mutagenesis is time-consuming and lacks specificity and precision due to the absence of adequate molecular information i.e., there is no opportunity to expand the gene pool, it is still a very practical approach, as it is quite simple and rapid method when compared to targeted mutagenesis.

Mutation procedures (either physical or chemical) can be optimized in term of the type of mutagen and dose. Mutagens specificity can be taken into account and mutagenesis itself can be enhanced or directed in order to obtain the maximum frequency of desirable mutant type among the isolates to be screened (**Mutwakil, 2011**).

Both physical (irradiation with ultraviolet) and chemical (treatment with NTG) mutagenesis techniques that have long been accepted as routine mutagenesis methods, therefore in the present study an attempt was carried out to improve the yield of citric acid from *A. niger* wild strains (FQW, KSUD and EMCC132) by UV and NTG mutation and grown under submerged fermentation conditions.

4.5.1. Physical mutation

Among the physical mutagens, ultraviolet radiation (UV) is often used to obtain hyper-producer strains (**Ali *et al.,* 2001**). The present study deals with screening and mutated the *A. niger* wild strains by UV irradiation for citric acid hyper-production under submerged fermentation condition.

For mutagenesis by UV, *A. niger* spores of wild strains were exposed to ultraviolet radiation at two wavelengths (254 and 365 nm) for three time intervals (60, 120 and 180 min). Then, fungal spores were grown in PD broth medium for 4-5 days at 30°C; Citric acid yield was assayed by HPLC (as mentioned for the sheets Figs. 7.6-7.9 in the appendix). Obtained data recorded in Table (4.V.1) and illustrated in Figure (4.V.1) indicate that the impact of UV radiation on citric acid productivity by irradiated *A. niger* strains was UV wave length dependent, where UV' wave length 254 nm was the most effective on all tested strains and their ability for citric acid production. Moreover, citric acid production by irradiated *A. niger* KSUD was increased to 32.16 mg/L (about three fold of unirradiated one) when exposed to UV (254 nm) for 60 min only.

Otherwise, UV' wave length 365 nm almost did not affect the strains, *A. niger* FQW and *A. niger* EMCC132, where their ability for citric acid production not significantly changed. It may due to genotype stability, reversal mutation or weakness of this wavelength in the events of mutations where it was close to visible light (**Sambrook and Russell 2001**).

In addition, irradiated *A. niger* KSUD was the most effected ones, it may be due to weakness or inability of the genetic repairing system and instability of genetic structure.

The obtained data revealed that hyper production of citric acid was showed in mutated strain compared to wild type strain. Among all the irradiated *A. niger* strains, it was observed that *A. niger* FQW gave highest amount of citric acid and researched to 41.98 mg/L after exposed to UV at 254 nm for 180 min.

The obtained results are in harmony with pervious data, where **Vasanthabharathi *et al.,* (2013)** revealed that hyper production of citric acid was observed in mutated strain compared to wild type strain. Among the mutated strain A2 which is exposed to UV for 4 hours was the most potential and it has produced 295.2 g/L citric acid. It was already found by **Ali *et al.,* (2001)** that 31.1 g/L citric acid was produced in parental strain whereas 50.0 g/l was produced by UV mutated strain.

Gupta and Sharma (1995) reported that the most frequently used method is induction by UV irradiation. Although the UV rays do not have much energy and, consequently, they do not induce ionization directly, the microorganisms are potent mutagenic agents, since UV rays are absorbed by purine and pyrimidines, making them reactive and inducing mutation **(Sambrook and Russell 2001)**. Therefore, the mutations induced by UV may randomly provide a strain with a higher capacity of citric acid production when compared with wild type strains citric acid production.

In addition, for strain mutation UV rays are important inducers. Main effect of this light is to modify the structure of pyrimidine (cytosine and thyamine) causing the formation of thyamine dimmer which distort the structure of DNA heliex and block the further replication process (**Sambrook and Russell 2001**). In most cases UV mutation are very harmful but at sometimes it may lead to better adaptation of an organism to its environment with improved biocatalytic performance.

Otherwise, referring to several previous researches, the hyper production may be also due to UV ray as a potent mutagen, where UV irradiation was proved to be best for the improvement of strains like *Aspergillus niger* for over production of various products when compared to the wild strains (**Kang** *et al.,* **1999; Ali** *et al.,* **2001; Soledad** *et al.,* **2006; Vasanthabharathi** *et al.,* **2013**).

Several investigators utilized UV irradiation for mutations induction from *A. niger* for citric acid production. The mutant strains might show several fold increase in citric acid production as compared to wild type cultures. **Hamissa** *et al.,* **(1992)** showed that UV treatment resulted in the development of 31 isolate of *A. niger* in citric acid production from beet molasses. **Rugsaseel** *et al.,* **(1993)** obtained mutants with enhanced citric acid production from soluble starch induced from *A. niger* WU-2223 after UV irradiation. **Lotfly** *et al.,* **(2007)** obtained UV mutant of *A. niger* in citric acid production approximately 2-3 fold increase when compared the parental wild type strain.

Table (4.V.1): Citric acid production (mg/l) by UV induced mutants of *A. niger* strains.

Fungal Strains	Radiations emission (nm)							
	254				365			
	Time of exposure (min)							
	0	60	120	180	0	60	120	180
FQW	13.918	13.923	40.907	41.978	13.918	14.561	14.250	14.190
KSUD	11.921	32.156	35.240	36.879	11.921	11.889	13.478	22.971
EMCC132	12.735	12.645	27.460	30.789	12.735	12.590	14.767	14.801

Spore suspensions of parental strains *A. niger* were subjected to mutagenesis with UV Radiations (254 and 365 nm); Time exposures (60, 120 and 180 min); grown in PD broth medium for 4-5 days at 30°C; Citric acid assay by HPLC.

A) Radiations emission 254 nm

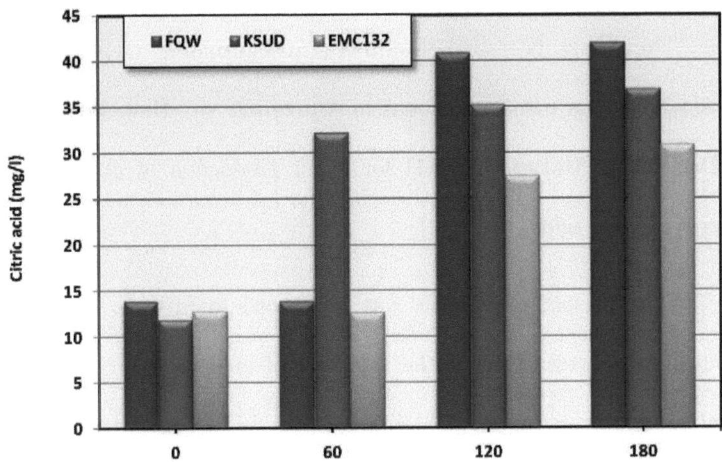

B) Radiations emission 365 nm

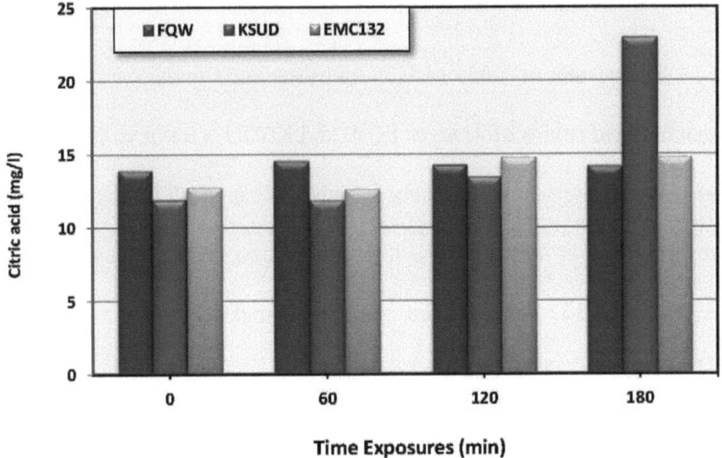

Fig. (4.V.1): Citric acid production (mg/l) by UV induced mutants of *A. niger* strains.

4.5.2. Chemical mutation

Previously N-methyl-N-nitro-N-nitrosoguanidine (NTG) has been widely used to induce mutations in *Aspergillus* sp. (**Haq, 2004; Pal and Das, 2005; Mutwakil, 2011**) for hyper production of certain valuable enzymes and acids.

The present experiment was performed to instigate the effect of chemical mutagen NTG in the induction of mutants in the conidia of *A. niger* wild strains (FQW, KSUD and EMCC132). Conidial spores of wild type *A. niger* treated with different concentrations of NTG 100, 200, 400, 600 and 750 mg/L) for different time exposures (30 and 60 min).

The results in Table (4.V.2) and illustrated in Figure (4.V.2) shows that mutated strains of *A. niger* FQW and KSUD were more faster response in terms of high citric acid concentration (22.6 and 27.7 mg citric acid/L, respectively) at 400 mg NTG/liter for 30 min. While muted strain of *A. niger* EMCC132 demonstrated more stable and affected at a concentration of MTG, 750 milligrams per liter for 30 min which is about 3 fold higher (36.2 mg citric acid/L) than its parental activity (12.6 mg citric acid/L).

When the wild strains of *A. niger* exposed to various concentrations of the chemical mutagen NTG for 60 min, the effectiveness was varied. Where citric acid production of wild strains was enhanced by mutated *A.*

niger FQW and KSUD to 28.4 and 26.7 mg citric acid/L at 200 and 100 µg NTG/ml, respectively. Otherwise, mutated *A. niger* EMCC132 was the most stable and showed about three times higher (35.6 mg citric acid/L) than the production of its parental strain (12.9 mg citric acid/L) at 400 mg NTG/liter for 60 min (Table 4.V.2).

In addition, among the citric acid productivity by mutated *A. niger* strains, it was found that exposure duration was directly proportional to the mutation rate by NTG which is successful method for efficient citric acid strains.

Previously, it has been well documented that N-methyl-N-nitrosoguanidine is one of the strong and multi-potential carcinogen that has been frequently reported inducing mutations (**Musilkova *et al.*, 1978; Zhu *et al.*, 2000; Pal and Das, 2005**).

NTG treated mutants provoked better excretion of citric acid in the fermented broth. **Haq (2004)** investigated that citric acid production by some selected mutant strains of *A. niger*. Among the 3 variants, *A. niger* GCM-45 was found to be a better producer of citric acid and it was further improved by chemical mutagenesis using *N*-methyl, *N*-nitro-*N*-nitrosoguanidine (MNNG).

Maximum yield was obtained from mutant type of *A. niger MTCC* 662 than wild strains were about 12.0g/100ml and 8.2g/100ml. It could be concluded that the *A. niger MTCC662* mutant strains have produced high yield than wild strain of *A. niger MTCC662* cultures (**Prasad *et al.,* 2014**).

Table (4.V.2): Citric acid production (mg/l) by NTG induced mutants of *A. niger* strains

Fungal strains	Time exposures (min)											
	30						60					
	NTG concentrations (μg/ml)											
	0	100	200	400	600	750	0	100	200	400	600	750
FQW	14.1	14.6	14.9	22.6	27.1	28.4	13.9	14.1	28.4	29.9	30.1	31.3
KSUD	12.3	12.6	12.9	27.7	30.2	31.2	12.0	26.7	26.8	27.4	27.6	27.8
EMCC132	12.6	13.8	14.1	14.3	14.5	36.2	12.9	13.1	12.95	35.6	35.9	36.1

Spore suspensions of parental strains *A. niger* were subjected to mutagenesis with NTG concentrations (0, 100, 200, 300, 400, 600 and 750 µg/ml); Time exposures (30 and 60 min); grown in PD broth medium for 4–5 days at 30°C; Citric acid assay by HPLC.

A) Time exposures 30 min

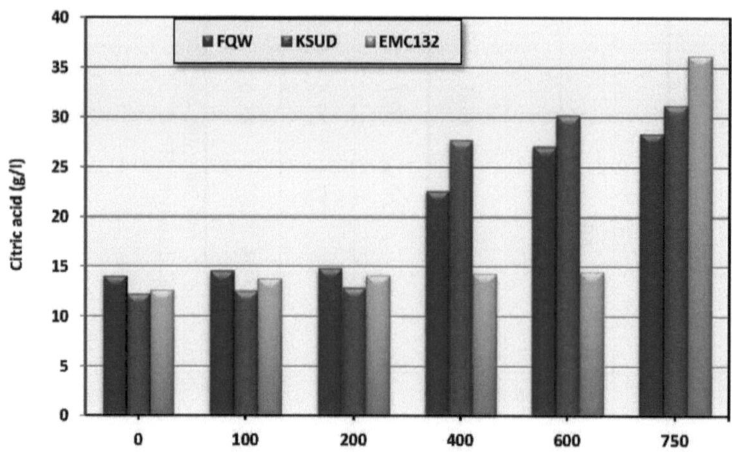

B) Time exposures 60 min

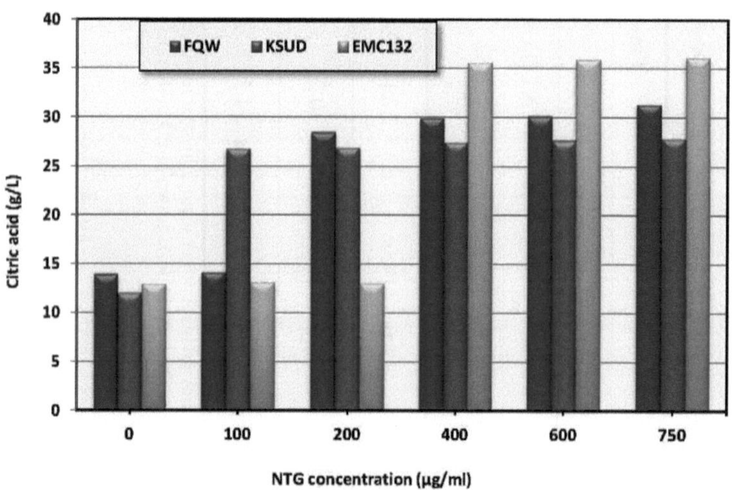

Fig. (4.V.2): Citric acid production (mg/l) by NTG induced mutants of *A. niger* strains

The number of reports considering strain improvement that have appeared in the literature is limited. Subsequently other perspectives for citric acid production sector are the improvement of citric acid producing strains, which have been carried out by mutagenesis and selection (**Soccol** *et al.,* **2006**). The most employed technique has been by inducing mutations in parental strains using mutagens. Mutants of *Aspergillus niger* are used for commercial production (**Haq** *et al.,* **2001**). Among mutagens, g-radiation, UV radiation and chemical mutagens are often used.

It could be concluded here that the treatment of fungal conidia with UV radiation and NTG resulted in the improved citric acid production. However, in 254 nm UV for 180 min developed mutant *A. niger* FQW (as a local strain) and the maximum citric acid production obtained (41.98 mg/L) were obtained (Table 4.V.1), which is three times higher than the original parental culture. On the other hand, mutagenized strain of *A. niger* FQW produced maximum yield of citric acid (31.3 mg/L) through treated with chemical mutagen NTG (750 µg/ml) for 60 min(Table 4.V.2 and Fig. 4.V.2), which is 2.25 folds higher than the original parental culture.

Mutant strain *A. niger* FQW with superior character as enhanced citric acid production is hope for investigators, where the most frequently used method for enhanced citric acid production is induction by UV irradiation.

Data obtained by **Rodrigues *et al.*, (2010)** in harmony with this result. Since UV rays are absorbed by Purine and Pyrimidines, making them reactive, inducing mutations (**Griffiths *et al.*, 2006 and Zaha, 2003**), therefore mutations induced by UV can randomly provide a strain with a higher capacity of citric acid production when compared to the parent strain for citric acid production (**Rodrigues *et al.*, 2010**)

The effectiveness of the induced mutagenesis depends on the type of mutagen, dose and duration of exposition (**Bapi *et al.*, 2004; Gochev and Krastanov 2007; Karanam and Medicherla 2008**). In the present study, although both mutagens UV-irradiation and NTG increased the ability of our strains to accumulate citric acid, but it was proved that physical mutagenesis by UV was more effective in comparison with chemical mutagenesis by NTG.

5. FUTURE PROSPECTS

On the basis of the obtained results, it can be recommended to:

1. Further improvements and optimization are required to develop the appropriate bioreactor and the fermentation parameters for citric acid production in semicontinuous or continues fermentation.

2. Study another an alternative cheap carbon source is needed to decrease the operational costs and shear in the reduction of environmental pollution.

3. Utilization the agro-industrial byproducts either in separate form or mixture of byproducts, can be used because they offer high levels of readily available carbon sources and another essential elements.

4. By product, cheese whey is a nutrient rich substrate and its high lactose content and may could utilize by *A. niger* for more efficiently citric acid production.

5. Further research is required for the biosafety of citric acid product before recommended for commercial utilization.

6. REFERENCES

Abou-Zeid A.A. and Ashy M.A. (1984). Production of citric acid: A review. Agr. wasters, 9:51-76

Adham N.Z. (2002). Attempts at improving citric acid fermentation by *Aspergillus niger* in beet-molasses medium. Bio resource Technology, 84:97–100.

Adinarayana K. and Prabhakar T. (2003). Optimisation of process parameters for cephalosporin C production under solid state fermentation from *Acremonium chrysogenum*. Process Biochemistry, 39(2):171–177.

Adinarayana K., Prabhakar T., Srinivasulu V., Anitha R.M., Jhansi L.P., and Ellaiah P. (2003). Optimization of process parameters for cephalosporin C production under solid state fermentation from *Acremonium chrysogenum*. Proc. Biochem., 39:171-177.

Ahmed S.A., Smith J.E. and Anderson J.A. (1972). Mitochondrial activity during citric acid production by *Aspergillus niger*. Trans Br. Mycol. Soc., 59:51–61.

Alam M.Z., Bari M.N., Muyibi S.A., Jamal P. and Abdullah M. (2011). Development of Culture Inoculum for Scale-Up Production of Citric Acid from Oil Palm Empty Fruit Bunches by *Aspergillus niger*. Procedia Environ. Sci., 8:396–402.

Ali S. (2004). Studies on the submerged fermentation of citric acid by *Aspergillus niger* in a stirred fermentor. Ph.D. Thesis. University of Punjab. Lahore. Pakistan, p. 59-60.

Ali S., Haq I. and Iqbal J. (2005). Effect of Low pH on continuous citric acid fermentation by *Aspergillus niger*. Pakistan Journal of Botany, 37(4):981-987.

Ali S., Haq I., Qadeer M.A. and Iqbal J. (2002). Production of citric acid by *Aspergillus niger* using cane molasses in a stirred fermentor. Electronic Journal of Biotechnology, 5(3):258-271.

Ali S., Haq I.U., Qadeer M.A. and Iqbal J. (2001). Biosynthesis of citric acid by locally isolated *Aspergillus niger* using sucrose salt media. J. Biol. Science, 1:178-181.

Alvarez-Vasquez F., Gonzalez-Alco C., and Torres N.V. (2000). Metabolism of citric acid production by *Aspergillus niger*: model definition, steady-state analysis and constrained optimization of citric acid production rate. Biotechnol. Bioeng., **70**:82-108.

Ambati P. and Ayyanna C. (2001). optimizing medium constituents and fermentation conditions for citric acid production from *Palmyra jaggery* using response surface method. Journal of Microbiology and Biotechnology, 17:331-335.

Anastassiadis S., Wandrey C. and Rehm H.-J. (2005). Continuous citric acid fermentation by *Candida oleophila* under nitrogen limitationat constant C/N ratio. World Journal of Microbiology & Biotechnology, 21:695–705.

Angumeenal A. R., Kamalakannan P., Prabhu H. J., and Venkappayya D. (2003a). Effect of transition metal cations on the production of citric acid using mixed cultures of *Aspergillus niger* & *Candida guilliermondii*. Journal of the Indian Chemical Society, 80:903-906.

Angumeenal A.R. and Venkappayya D. (2005a). Effect of transition metal ions on the metabolism of *Aspergillus niger* in the production of citric acid with molasses as substrate. Journal of Scientific and Industrial Research, 64:125-128.

Angumeenal A.R. and Venkappayya D. (2005b). Artrocarpus heterophyllus e a potential substrate for citric acid biosyntheis using *Aspergillus niger*. LWT e Food Science and Technology, 38:89-93.

Angumeenal A.R. and Venkappayya D. (2005c). Bioconversion of Amorphophallus campanulatus to citric acid by *Aspergillus niger* e effect of metal ions on fermentation, modeling studies and correlarion of theoretical and experimental parameters. Indian Journal of Biotechnology, 4:246-250.

Angumeenal A.R., Kamalakannan P., Prabhu H.J. and Venkappayya D. (2003b). Bioconversion of Colocassia antiquorum and Aponogetannatans to citric acid by *Aspergillus niger* e effect of metal ions and kinetics. Journal of Scientific and Industrial Research, 62:447-452.

Anwar S., Ali S. and Sardar A.A. (2009). Citric acid fermentation of hydrolyzed raw starch by *Aspergillus niger* IIB-A6 in stationary culture. Sindh University Research Journal, 41(1):01-08.

Auta H. S., Abidoye K.T., Tahir H., Ibrahim A.D. and Aransiola S.A. (2014). Citric Acid Production by *Aspergillus niger* Cultivated on *Parkia* *biglobosa* Fruit Pulp. International Scholarly Research Notices, p. 1-8. Article ID 762021. http://dx.doi.org/10.1155/2014/762021.

Baby, E. A., & Saad, M. (1996). Production of citric acid by a heavy metal adapted *Aspergillus niger* NRRL 595. African Journal of Mycology and Biotechnology, 4, 59e69.

Bapi Raju K.V.V.S.N., Sujatha P., Ellaiah P. and Ramana T. (2004). Mutation induced enhanced biosynthesis of lipase. African J. Biotechnol., 3:618-621.

Bari M.N., Alam M.Z., Muyibi S.A., Jamal P. and Mamun A. (2009). Improvement of production of citric acid from oil palm empty fruit bunches: Optimization of media by statistical experimental designs Bioresource Technol., 100: 3113-3120.

Barrington S. and Kim J.-W. (2008). Response surface optimization of medium components for citric acid production by *Aspergillus niger* NRRL 567 grown in peat moss. Bio resource Technology, 99:368–377.

Bonatelli Jr, R. and Azevedo, J. L, (1983). Improved reproducibility of citric acid production in *Aspergillus niger*. *Biotechnol.Lett.*, **4**, 761-766

Boominathan S., Dhanaraj T.S. and Murugaiah K. (2012). Production and optimization of citric acid by Aspergillus niger using pineapple and tapioca waste. Herbal Tech. Industry, 1(5):8-11.

Chaudary K., Ethiraj S., Lakshminarayana K. and Tauro P. (1978). Citric acid production from Indian cane molasses by *Aspergillus niger* under solid state fermentation condition. Journal of Research Haryana Agricultural University, 1:48–52.

Currie J.N. (1917). The citric acid fermentation of *A. niger*. J. Biol. Chem., 31:5.

Darani K. and Zoghi A. (2008). Comparison of pretreatment strategies of sugarcane bagasse: Experimental design for citric acid production. Bioresource Technology, 99: 6986- 6993.

Das, A. and Roy, P. (1978). Improved production of citric acid by diploid strain of *Aspergillus niger*. Can. J. Micobiol., **24**, 622-625

Dasgupta J., Nasim S., Khan A.W. and Vora V.C. (1994). Production of citric acid in molasses medium: effect of addition of lower alcohols during fermentation, J. Microbiol. Biotechnol., 9:123–125.

Demirel G., Yaykali K.O. and Yaar A. (2005). The production of citric acid by using immobilized *Aspergillus niger* A-9 and investigation of its various effects. Food Chemistry, 89(3):393-396.

Dhillon G.S., Brar S.K., Kaur S. and Verma M. (2013). Bioproduction and extraction optimization of citric acid from *Aspergillus niger* by rotating drum type solid-state bioreactor. Ind. Crops Prod., 41:78–84.

El-Aasar S.A. (2006). Submerged Fermentation of Cheese Whey and Molasses for Citric acid Production by *Aspergillus niger*. International Journal of Agriculture & Biology, 4:463–467.

EL-Holi M.A. and AL-Delaimy K.S. (2003). Citric acid production from whey with sugars and additives by *Aspergillus niger*. African J. Biotechnol., 2:356–259.

El-Hussein A.A., Tawfig S.A.M., Mohammed S.G., El-Siddig M.A. and Siddig M.A.M. (2009). Citric acid production from kenana cane molasses by *Aspergillus niger* in submerged fermentation. Journal of Genetic Engineering and Biotechnology, 7(2):51-57.

Ellaiah P., Srinivasalu B. and Adinarayana K. (2003). Optimisation studies on neomycin production by a mutant strain of **Streptomyces marinensis** in solid state fermentation. Process Biochemistry, 39, 529–534.

Faruk M.O., Mumtaz T. and Rashid H. (2014). Supplementary effect of potato peel htdrolysate on the citric acid production by Aspergillus niger CA16. Biotechnology, 9(8):308-310.

Favela-Torres E., Cordova-Lopez J., Garcia-River, M. and Gutierrez-Rojas M. (1998). Kinetics of growth of *Aspergillus niger* during submerged, agar surface and solid state fermentations. Proc. Biochem., 33: 103-107.

Fawole O.B. and Odunfa S.A. (2003). Some factors affecting production of pectin enzymes by *Aspergillus niger*. Int. Biodet. Biodeg., 51:223-227.

Fawole O.B. and Odunfa S.A. (2003). Some factors affecting production of pectic enzymes by *Aspergillus niger*. International Biodeterioration and Biodegradation, 52(4), 223–227.

Femi-Ola T.O. and Atere V.A. (2013). Citric acid production from brewers spent grain by *Aspergillus niger* and *Saccharomyces cerevisiae*. International Journal of Research in BioSciences, 2:30–36.

Garg, N., & Hang, Y. D. (1995). Microbial production of organic acids from carrot processing waste. Journal of Food Science and Technology, 32, 119e121.

Gochev V. and Krastanov A. (2007). Enhanced Production of Extracellular Lipase by Novel Mutant Strain of *Aspergillus niger*. Ecology and Safety Int. Sci. Publ., 1, 260-267.

Grewal H.S. and Kalra K.L. (1995). Fungal production of citric acid. Biotechnology Advances, 13 (2):209-234.

Griffiths A.J.F., Wesller S.R., Lewontin R.C., Gelbart W.H., Suzuki D.T. and Miller J.H. (2006). Introducao a genetica (8th ed.) Rio de Janeiro-RJ, Brazil: Guanbara- Koogan.

Grimoux E. and Adams P. (1880). Synthese de lácide citrique. C R Hebd Seances Acad. Sci., 90:1252. C.F Papagianni M. (2007).

Gunde-Cimerman, N., Cimerman, A., Perdhi, A.(1986) *Aspergillus niger* mutants for bioconversion of apple distillery wastes.Enzyme Microb. Technol., 8, 166-170.

Gupta J.K., Heding L.G. and Jorgensen O.B. (1976). Effect of sugars, hydrogen ion concentration and ammonium nitrate on the formation of citric acid by *Aspergillus niger*. Acta Microbiol. Acad. Sci. Hung., 23:63– 67.

Gupta S. and Sharma C.B. (1995). Citric acid fermentation by mutant strain of *Aspergillus niger* resistant to manganese ions inhibition. Biotechnol. Letters, 17:269-274.

Hamissa F.A., El-Abyad A.M., Abdu A. and Gad S.A. (1992). Raising potent UV mutants of *Aspergillus niger* van Tieghem for citric acid production from beet molasses. Bioresource Technology, 39:209-213.

Hang Y.D. and Woodams E.E. (1998). Production of citric acid from corncobs by *Aspergillus niger*. Biores. Technol., 65:251-253.

Hang Y.D., Splittstoesser D.F., Woodams E.E. and Sherman R.M. (1977). Citric acid fermentation of brewery waste. J. Food Sci., 42:383-384.

Hang, Y. D., & Woodams, E. E. (1984). Apple pomace; a potential substrate for citric acid production by *Aspergillus niger*. Biotechnology Letters, 6, 763e764.

Hang, Y. D., & Woodams, E. E. (1987). Microbial production of citric acid by solid state fermentation of kiwi fruit peel. Journal of Food Science, 52, 226e227.

Hang, Y. D., &Woodams, E. E. (1985). Grape pomace; a novel substrate for microbial production of citric acid. Biotechnology Letters, 7, 253e254

Haq I., Ali S., Qadeer M.A. and Iqbal J. (2002). Citric acid fermentation by mutant strain of *Aspergillus niger* GCMC-7 using molasses based medium. Electronic Journal of Biotechnology, 5(2):125-132.

Haq I., Ali S., Qadeer M.A. and Iqbal J. (2004). Citric acid production by selected mutants of *Aspergillus niger* from cane molasses. Bioresourse Technology, 93(2):125-130.

Haq I., Ali S., Qadeer M.A. and Javed I. (2003). Stimulatory effect of alcohols (methanol and ethanol) on citric acid productivity by a 2-deoxy

D-glucose resistant culture of *Aspergillus niger* GCB-47. Bioresource Technol., 86:227-233.

Haq I., Khurshid S., Ali S., Ashraf H., Qadeer M.A. and Rajoka M.I. (2001). Mutation of *Aspergillus niger* for hyperproduction of citric acid from black strap molasses. World J. Microbiol. Biotechnol., 17:35–37.

Honecker S., Bisping B., Yang Z. and Rehm H.J. (1989). Influence of sucrose concentration and phosphate limitation on citric acid production by immobilized cells of *Aspergillus niger*. Appl. Microbiol. Biotechnol., 31:17–24.

Hossain M., Brooks J.D. and Maddox I.S. (1984). The effect of the sugar source on citric acid production by *Aspergillus niger*. Appl. Microb. Biotechnol, 19:393-397.

Ikram-ul-Hag; Samina K; Sikander A.; Hamed A.; Qadeer M.A. and Ibrahim R.M. (2001). Mutation of Aspergillus niger for hyperproduction of citric acid from black strap molasses. World J. of Microbiology and Biotechnology, 17: 35-37.

Ingram L.O. and Buttke T.M. (1984). Effects of alcohols on microorganisms. Adv. Microb. Physiol., 25:253.

Islam, M. S., Begum, R. and Choudhury, N.(1986) Semipilot scale production of citric acid in cane molasses by gamma ray induced mutant of *Aspergillus niger*. Enzyme Microb. Technol., **8**, 461-471

Jana A.K. and Ghosh P. (1995). Xantan biosynthesis in continuous culture: Citric acid as an energy source. J. Ferm. Bioeng., 80:485-491

Jernejc K., Perdih A. and Cimerman A. (1992). Biochemical composition of Aspergillus niger mycelium grown in citric acid productive and non-productive conditions, J. Biotechnol., 25:341–348.

Jianlong W. and Ping L. (1998). Phytate as a stimulator of citric acid production by *Aspergillus niger*. Proc. Biochem., 33:313-316.

Kang S.W., Ho H.E., Lee J.S. and Kim S.W. (1999). Overproduction of beta glucosidase by *Aspergillus niger* mutant from lignocellulosic biomass. Biotechnol. Lett.; 21: 647-650.

Kapelli O., Muller M. and Fiechter A. (1978). Chemical and structural alterations at cell surface of *Candida tropicalis* induced by hydrocarbon substrate. Journal of Bacteriology, 133:952-958.

Karanam S.K. and Medicherla N.R. (2008). Enhanced lipase production by mutation induced *Aspergillus japonicas*. African J. Biotechnol., 7:2064-2067.

Kareem S.O. and Rahman R.A. (2013). Utilization of banana peels for citric acid production by *Aspergillus niger*. Agriculture and Biology Journal of North America, 4(4):384–387.

Kareem S.O., Akpan I. and Alebiowu O.O. (2010). Production of citric acid by *Aspergillus niger* using pineapple waste. Malaysian Journal of Microbiology, 6(2):161–165.

Karthikeyan A. and Sivakumar N. (2010). Citric acid production by Koji fermentation using banana peel as a novel substrate. Bioresource Technology, 101(14):5552–5556.

Kim J.W., Barrington S. and Lee B.H. (2004). Cheese whey for citric acid production by *Aspergillus niger* NRRL 567 under submerged fermentation. Part I. Optimizing nutrient supplementation. Appl. Microbiol. Biotechnol., 92(1):97-101.

Kirimura K, Honda Y, and Hattori T. (2011). Citric Acid. Comp. Biotechnol., 3:135–142.

Kirimura, K., Lee, S. P. Kawajima, I., Kawabe,S. and Usami, S. (1988a) Improvement in citric acid production by haploidization of *Aspergillus niger* diploid strains. *J. Ferment. Technol.*, **66**, 375-382

Kishore K.A. and Reddy G.V. (2012). Optimization of incubation time, fermentation temperature & O_2 flow rate in citric acid fermentation using response surface methodology (RSM). IJAET (International Journal of Advanced Engineering Technology), III(1):64-67.

Kishore K.A., Kumar M.P., Krishna V.R., Ravi V. and Reddy G.V. (2008). Optimization of process variables of citric acid production using *Aspergillus niger* in a batch fermentor. Engineering Letters., 16 (4):572.

Kovats J. (1960). Studies on submerged citric acid fermentation. Acta Microbiol., 28:26-33.

Krishna C. (2005). Solid-State Fermentation Systems - An Overview. Critical Reviews in Biotechnology, 25::1– 30.

Kristiansen B. and Sinclair C.G. (1978). Production of citric acid in batch culture. Biotechnol. Bioeng., 20:1711–1722.

Kubicek C.P. and Rohr M. (1986). Citric acid fermentation, CRC Crit. Rev. Biotechnol., 3:331-373.

Kumar D., Jain V.K., Shanker G., and Srivastava A. (2003). Utilisation of fruits waste for citric acid production by solid state fermentation. Proc. Biochem., 38:1725-1729

Kundu, S., Panda, T., Majumdar, S. K., Guha, B., & Bandyopadhyay, K. K. (1984). Pretreatment of Indian cane molasses for increased production of citric acid. Biotechnology and Bioengineering, 26, 1114e1121.

Leangon S., Maddox I.S. and Brooks J.D. (2000). A proposed biochemical mechanism for citric acid accumulation by *Aspergillus niger* Yang No.2 growing in solid state fermentation. W. J. Microbiol. Biotechnol., 16:271-275.

Lotfy W.A., Ghanem K. M., and El-Helow E. R. (2007a). Citric acid production by a novel *Aspergillus niger* isolate: II. Optimization of process parameters through statistical experimental designs. *Bioresource Technology*, vol. 98, no. 18, pp. 3470–3477.

Lotfy W.A.; Khaled M. G. and Ehab R.E. (2007b). Citric acid production by a novel *Aspergillus niger* isolate I. Mutagenesis and cost reduction studies. Bioresource Technology‹ 98:3464-3469.

Lu, Brooks, M. J. D., & Maddox, I. S. (1997). Citric acid production by solid state fermentation in a packed bed reactor using *Aspergillus niger*. Enzyme Microbial Technology, 21, 392-397.

Maddox I. S., Spencer K., Greenwood J. M., Dawson M. W. and Brooks J. D. (1985). Production of citric acid from sugars present in wood hemicellulose using *Aspergillus niger* and *Saccharomycopsis lipolytica*. Biotechnol. Lett., 7:815-818.

Maharani V., Reeta D., Sundaramanickam A., Vijayalakshmi S. and Balasubramanian T. (2014). Isolation and characterization of citric acid producing *Aspergillus niger* from spoiled coconut. Int. J. Curr. Microbiol. App. Sci., 3(3):700-705.

Majumdar, L., Khalid, I., Munshi, M. K., Alam, K., & Or-Rashid, H. (2010). Citric acid production by *Aspergillus niger* using molasses and pumpkin as substrates.European Journal of Biological Sciences, 2, 1e8.

Majumder L., Khalil M.I, Munshi M.K., Alam M.K., Rashid H.O. and Alam N. (2009). Gamma-Ray Induced Mutants 14/20 and 79/20 of *Aspergillus niger* Increases Citric Acid Production from Molasses and Pumpkin. African Journal of Basic & Applied Sciences, 1(1-2): 26-30.

Marier J.R. and Boulet M. (1958). Direct determination of citric acid in milk with improved pyridine-acetic anhydride method. J. Dairy Sci., 41:1683-1692.

Mattey M. (1992). The production of organic acids. Crit. Rev. Biotechnol., 12:87–132.

Max B., Salgado J.M., Rodríguez N., Cortés S., Converti A. and Domíngue J.M. (2010). Biotechnological production of citric acid. Brazilian Journal of Microbiology, 41:862-875.

Meixner O., Mischak H., Kubicek C.P. and Roh M. (1985). Effect of manganese deficiency on plasma membrane lipid composition and glucose uptake in Aspergillus niger. FEMS Microbiol. Lett., 26:271-275.

Milson P.E. (1987). Organic acids by fermentation, especially citric acid. Food Biotechnology. 1st Edited by King R.D. and Cheetam P.S.J., Applied Science, London and New York, pp. 273-306.

Mirminachi F., Zhang A. and Roehr M. (2002). Citric acid fermentation and heavy metal ions: Effect of iron, manganese and copper. Acta Biotechnol., 22:363-373.

Mitard A. and Riba A. (1988). Morphology and growth of *Aspergillus niger* ATCC 26036 cultivated at several shear rates. Biotechnol. Bioeng., 32:835-840.

Moataza, M. S. (2006). Citric acid production from pretreated crude date syrup by *Aspergillus niger* NRRL595. Journal of Applied Sciences Research, 2, 74e79

Moore-Landecker E. (1996). Fundamentals of the fungi. Fourth Edition. Prentice-Hall Inc., Upper Saddle River, New Jersey, USA. Pp. 574.

Musilkova, M., Ujcova, E., Seichert, L, Fencl, Z.(1983) Effect of changed cultivation conditions o the morphology of *Aspergillus niger* and citric acid biosynthesis in laboratory cultivation. *Folia Microbiol.,* **27,**382 332

Musilkova, M., Z. Fencl, E. Ujcova and L. Seichert, (1978). Variability of *Aspergillus niger* after treatment with N-methyl-N-nitro-Nnitrosoguanidine., Flia Microbio., 23(2):103-107.

Mutwakil M.H.Z., (2011). Mutation induction in *Aspergillus terrus* using N-methyl-N'-nitro-N-nitrosoguanidine (NTG) and Gamma rays. Australian Journal of Basic and Applied Sciences, 5(12):496-500.

Nadeem A., Syed Q., Baig S., Irfan M. and Nadeem M. (2010). Enhanced Production of Citric Acid by *Aspergillus niger* M-101 Using Lower Alcohols. Turkish Journal of Biochemistry, 35(1):7-13.

Nampoothiri M., Baiju T.V., Sandhya C., Sabu A., George S., Ashok P. (2004). Process optimization for antifungal chitinase production by *Trichoderma harzianum*. Process Biochem., 39:1583–1590.

Nason A. (1968). Essentials of modern biology. John Wiley and Sons Inc.,m New York. Pp. 373-376.

Navaratnam P., Arasaratnam, V. and Balasubramaniam, K. (1998). Channeling of glucose by methanol for citric acid production from *Aspergillus niger*. W. J. Microbiol. Biotechnol., **14**:559-563

Nour V., Trandafir I. and Ionica M. E. (2010). HPLC Organic Acid Analysis in Different Citrus Juices under Reversed Phase Conditions. Not. Bot. Hort. Agrobot. Cluj., 38(1):44-48.

Orthofer R., Kubicek C.P. and Rohr M. (1979). Lipid levels and manganese deficiency in citric acid producing strains of Aspergillus niger. FEMS Microbiol. Lett., 5:403.

Pal, S.K. and T.K. Das, 2005. Biochemical characterization of N-methyl N-nitro-N-nitrosoguanidine induced cadmium resistant mutants of *Aspergillus niger*. J. Biosci., 30:639-646.

Pallares J., Rodriguez S. and Sanroman A. (1996). Citric acid production in submerged and solid state culture of *Aspergillus niger*. Bioprocess Eng., 15:31–33.

Panda.T, Kundu.S, Majumdar. SK.(1984) Studies on **citric acid** production by *Aspergillus niger* using treated cane molasses. Proc. Biochem., 19, pp. 183–187.

Pandey A., Soccol C.R., Rodriguez-Leon J.A. and Nigam P. (2001). Production of Organic Acids by Solid-State Fermentation. In: Solid-State Fermentation in Biotechnology – Fundamentals and Applications, Asiatech Publishers Inc., New Delhi, India. pp. 113–126.

Pandey P., S. Putatunda, L. Dewangan, V. S. Pawar, and S. A. Belorkar, (2013). Studies on citric acid production by *Aspergillus niger* in batch fermentation. Recent Research in Science and Technology, vol. 5, no. 2, pp. 66–67.

Papagianni M, Mattey M, Berovic M and Kristiansen B (1999). *Aspergillus niger* morphology and citric acid production in submerged batch fermentation: effects of culture pH, phosphate and manganese levels. Food Technol Biotechnol., 37:165–71.

Papagianni M. (2004). Fungal morphology and metabolite production in submerged mycelial processes. Biotechnology Advances, 22(3):189-259.

Papagianni M. (2007). Advances in citric acid fermentation by *Aspergillus niger*: Biochemical aspects, membrane transport and modeling. Biotechnology Advances, 25:244–263.

Papagianni M. and Mattey M. (2006). Morphological development of Aspergillus niger in submerged citric acid fermentation as a function of the spore inoculum level. Application of neural network and cluster analysis for characterization of fungal morphology. Microbial Cell Factories, 5:3.

Papagianni M., Mattey M. and Kristiansen B. (1998). Citric acid production and morphology of *Aspergillus niger* as functions of the mixing intensity in a stirred tank and a tubular loop bioreactor. Biochemical Engineering Journal, 2:197-205.

Papagianni M., Mattey M., and Kristiansen B. (1994). Morphology and citric acid production of *Aspergillus niger PM 1*. Biotechnol. Lett., 16(9):929-934.

Partos L. (2005). ADM closes citric acid plant as Chinese competition bites, Food Navigator. http://www.foodnavigator.com/Market-Trends/ADM-closes-citric-acid-plant-as-Chinese-competition-bites.

Paul G.C., Priede M.A. and Thomas C.R. (1999). Relationship between morphology and citric acid production in submerged *Aspergillus niger* fermentations. Biochemical Engineering Journal, 3(2):121-129.

Pazouki, M., Felse, P.A., Sinha, J., Panda, T. (2000). Comparative studies on citric acid production by *Aspergillus niger* and *Candida lipolytica* using molasses and glucose. Bioproc. Eng., **22**:353-361.

Pelechova, J., Petrova, L., Ujcova, E. and Martinkova, L. (1990) Selection of ahyperproducing strain of *Aspergillus niger* for biosynthesis of citric acid on unusual carbon substrates. *Folia Microbiol.*, **35**, 138-142

Penniston K.L., Nakada S.Y., Holmes R.P. and Assimos D.G. (2008). Quantitative assessment of citric acid in lemon juice, lime juice, and commercially-available fruit juice products. Journal of Endourology, 22(3):567–570.

Pessoa D.F.; Diasde C. and Angela C. (1982). Production of citric acid by *Aspergillus niger*. Revista de Microbiologia, 13(2):225-229.

Pessoa F.F., Castro A.C. and Leite S.G. (1984). Citric acid fermentation with *Aspergillus niger*. Reviews in Microbiology, 15:89-93.

PintadoJ., Torrado A., Gonzalez M.P., Murado M.A.(1998) Optimizationof nutrient concentration for citric acid production by solid-state culture of *Aspergillus niger* on polyurethane foams, *Enzyme Microb. Technol. 23* 149–156.

Pontecorvo, G., Roper, J. A. and Forbes, E.(1953a) Genetic recombination without sexual reproduction in *Aspergillus niger*.J.Gen. Microbiol., **8**,198

Prasad M.P.D., Surendra Babu N.V., Sridevi V., Reddy O.V.S, Prakasam R.S and (2014). Response surface methodology for optimization of sorghum malt media for citric acid production by improved strains of *Aspergillus niger*. International Journal of Advanced Research, 2:498-507.

Rane K.D. and Sims K.A. (1993). Production of citric acid by *Candida lipolytica*, Y-1095; effect of glucose concentration on yield and productivity. Enzyme Microbial Technology, 15:646-651.

Rane KD and Sims KA. (1996). Citric acid production by *Yarrowia lipolytica*: effect of nitrogen and biomass concentration on yield and productivity. Biotechnology Letters.;18(10):1139-1144.

Rodrigues C.; Vandenberghe.; L.P.; Teodoro J.; Pandey A. and Soccol C.R. (2010): Improvement on citric acid production in solid. State fermentation by *Aspergillus niger* LPBBC mutant using citric pulp. Appl. Biochem. Biotechnol., 8(8): 8370-8375.

Rojoka M.I.; Ahmed M.N.; Shahid F.L. and Pervez S. (1998). Citric acid production from sugarcane molasses by Aspergillus niger. Biologia., 44: 241-253.

Roukas T. (1991). Production of citric acid from beet molasses by immobilized cells of *Aspergillus niger*. Journal of Food Science, 56:878–880.

Roukas T. (1998). Carob pod, A new substrate for citric acid production by *Aspergillus niger*. Applied Biochemistry and Biotechnology, 74, 43e53.

Roukas, T. and Kotzekidou, P. (1997). Pretreatment of date syrup to increase citric acid production. Enzyme and Microbial Technology, 21, 273e276.

Rugsaseel S., Kirimura K. and Usami S. (1993). Selection of mutants of *Aspergillus niger* showing enhanced productivity of citric acid from starch in shaking culture. J. of Fermentation and Bioengineering, 75:226-228.

Sakurai A., Gant H. and Hurochi I. (1996). Effect of oxygen tension on citric acid production by surface culture of *Aspergillus niger*. J. Chem. Technol. And Biotechnol., 65:432-437.

Sambrook J. and Russell D.W. 2001. Molecular cloning: a laboratory manual, 3rd edition. Cold Spring Harbor Laboratory Press, Cold Spring Harbor.

Sanat E.I.S. and Korish M. (2007). Utilisation of potato chips waste for production of high value products: production of citric acid by solid state fermentation. Alex Journal of Food Science & Technology, 4, 19e26.

Sanjay K. and Sharma P. (1994). A high performance process for production of citric acid from sugarcane molasses. J. Microbiol., 23:211-217.

Sassi G., Ruggeri B., Specchia V., and Gianetto A. (1991). Citric acid production by *Aspergillus niger* with banana extract. Bioresource Technology, 37, 259-269.

Sayer J.A. and Gadd G.M. (2001). Binding of cobalt and zinc by oragnic acids and culture filterates of *Aspergillus niger* grown in the absence or presence of insoluble cobalt or zinc phosphate. Mycological Research, 105:1261-1267.

Schuster E., Dunn-Colema, N., Frisva, J C. and Van Dije, P W. (2002). On the safety of *Aspergillus niger*–A review. Applied Microbiology Biotechnology, 59:426-435.

Shadafza D., Ogawa T. and Fazeli A. (1976). Comparison of citric acid production from beet molasses and date syrup with *Aspergillus niger*. Hakko Kogaku Zasshi, 54:65-75.

Shankaran V. S. and Lonsane B.K. (1994). Coffee husk and inexpensive substrate for production of citric acid by *Aspergillus niger* in a solid state fermentation. World Journal of Microbiology and Biotechnology, 10, 165-168.

Shankaranand V.S. and Lonsane B.K. (1994). Ability of Aspergillus niger to tolerate metal ions and minerals in a solid-sate fermentation system for the production of citric acid. Process Biochem., 29:29-37.

Shu P. and Johnson M.J. (1948). The interdependence of medium constituents in citric acid production by submerged fermentation. J. Bacteriol., 54:161–167.

Smith J.J., Lilly M.D. and Fox R.I. (1990). Biotechnol. Bioeng., 35:1011-1023.

Soccol C.R., Vandenberghe L.P.S, Rodrigues C. and Pandey A. (2006). New Perspectives for Citric Acid Production and Application. Food Technol. Biotechnol., 44(2):141–149.

Soccol C.R. and Vandenberghe L.P.S. (2003). Overview of applied solid-state fermentation in Brazil, Biochem. Eng. J. 13205–218.

Soledad D.N.S., González C.R., Almendárez B.G., Francisco J. and Fernández A.T. and Sergio H. (2006). Physiological, morphological, and mannanase production studies on *Aspergillus niger* UAM-GS1 mutants. Electronic J. Biotechnol., 9:10-19.

Sukesh K, Jayasuni J.S, Gokul C.N and Anu V (2013). Citric acid production from agronomic waste using *Aspergillus niger* isolated from decayed fruit. Journal of Chemical, Biological and Physical Sciences, 3(2):1572-1576.

Suzuki A.,Sarangbin S.,Kirimura K. and Usami S.(1996). Direct Production of Citric Acid from Starch by a 2-Deoxyglucose- Resistant Mutant Strain of *Aspergillus niger*. Journal of fermentation and bioengineering, 81:320-323,

Tran C.T. and LiSly, D.A. (1998). Mitchel, selection of a strain of *Aspergillus niger* for the production of citric acid from pine apple waste in solid state fermentation.World Journal of Microbiology and Biotechnology, 14, 399-404.

Ujcova E., Fencl Z., Musilcova M. and Scichert L. (1980). Morphology and citric acid production of *Aspergillus niger*. Biotechnology and Bioengineering, 22:237-241.

Usha T., Venkateshwaran G., Sarada R. and Ravishankar G.A. (2001). Studies on *Haematococcus pluvialis* for improved production of astaxanthin by mutagenesis. World J. Microbiol. Biotechnol., 17:143-148.

Vaija J. and Linko P. (1986). Continuous citric acid production by immobilized *Aspergillus niger*: reactor performance and fermentation techniques. Journal of Molecular Catalysis, 38:237-253.

Vaishnavi R., Chairman K., Ranjit Singh A.J.A., Ramesh S., and Viswanathan S. (2012). Screening the effect of bagasse -an agrowaste for the production of citric acid using *Aspergillus niger* through solid state fermentation. Bioscience Methods, 3(8):48–54.

Van Suijdam J.C., Kossen N.W.F. and Paul P.G. (1980). An inoculums technique for the production of fungal pellets. European Journal of Applied Microbiology and Biotechnology, 10(3):211–221.

Vandenberghe L.P.S. (2000). Development of process for citric acid production by solid-state fermentation using cassava agro-industrial residues, PhD Thesis, Université de Technologie de Compiègne, Compiègne, France. pp. 205.

Vandenberghe L.P.S., Soccol C.R., Pandey A. and Lebeault J.M. (1999). Review: Microbial production of citric acid, Braz. Arch. Biol. Technol., 42:263–276.

Vandenberghe L.P.S., Soccol. C.R., Prado F.C., Pandey A. (2004) Comparison of citric acid production by solid-state fermentation in flask, column, tray and drum bioreactor, Appl. Biochem. Biotechnol. 118:1–10.

Vasanthabharathi V., Sajitha N. and Jayalakshmi S. (2013). Citric Acid Production from U-V Mutated Estuarine *Aspergillus niger*. Advances in Biological Research, 7 (3):89-94.

Vergano M.G.F., Soria M.A. and Kerber N.L. (1996). Influence of inoculums preparation on citric acid production by *Aspergillus niger*. World J. Microbiol. Biotechnol., 12:655–656.

Wang J.L. (1998). Improvement of citric acid production by *Aspergillus niger* with addition of phytate to beet molasses. Bioresource Technology, 65(3):243-245.

Wayman F.M. and Mattey M. (2000). Simple diffusion is the primary mechanism for glucose uptake during the production phase of the *Aspergillus niger* citric acid process. Biotechnology and Bioengineering, 67(4):451-456.

Wehmer C. (1893). Note sur la fermentation citrique. Bull. Soc. Chem., 9:728. C.F.: Max *et al.*, (2010).

Wojtatowics M., Marchin G.L. and Ericksen L.E. (1993). Attempts to improve strain A-10 of *Yarrow lipolytica* for citric acid production from n-parrafins. Process Biochemistry, 28:453-460.

Xu D.-B., Madrid C. P., Rohr M. and Kubicek C. P. (1989). The influence of type and concentration of the carbon source on production of citric acid by *Aspergillus niger*. Appl. Microbiol. Biotechnol., 30:553-558.

Yadegary M., Hamidi A., Alavi S.A., Khodaverdi E., Yahaghi H., Sattari S., Bagherpour G. and Yahaghi E. (2013). Citric acid production from sugarcane bagasse through solid state fermentation method using *aspergillus niger* mold and optimization of citric acid production by taguchi method. Jundishapur J. Microbiol., 6(9):7625.

Yokoya F. (1992). Citric Acid Production. In: Industrial Fermentation Series, Campinas, SP, Brazil pp. 1–82.

Zafiris A., Tzia G., Orea C., Poulou V. and Thomopoulos C.D. (1994). Fermentation of orange processing waste for citric acid production. Journal of Science of Food and Agriculture, 65, 117-120.

Zaha, A. (2003): Biologia molecular basica (3rd ed.) Porto Alegre-RS, Barazil: Mercado Aberto.

Zhu, J.S., Q. Su, J.G. Zhau, P.L. Hi and J.H. Xo, (2000). Study of primary leiomyosarcoma induced by MNNG in BALB/C nude mice. World J. Gasterointrol., 6(1):128-130.

7. Appendix

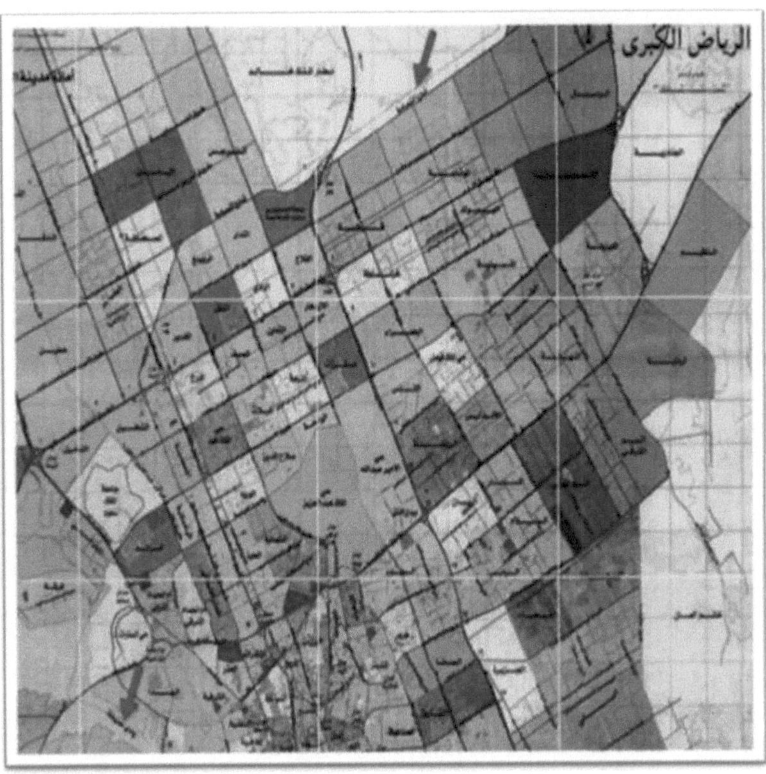

Fig. (7.1): Map of Riyadh region showing the position of soil samples sites, Thumama and Wadi Hanifa.

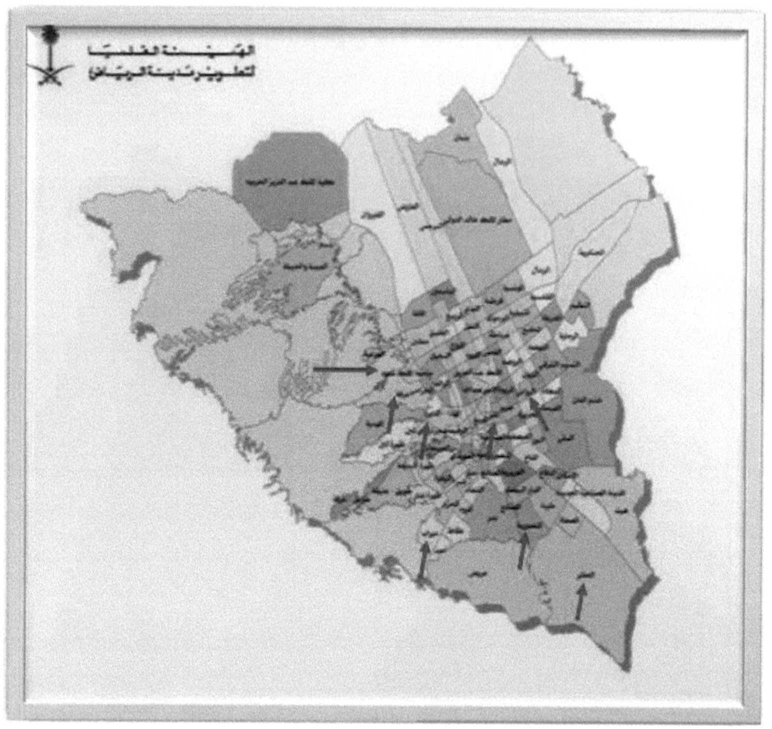

Fig. (7.2): Map of Riyadh city showing the position of soil samples sites, Malaz, Rawabi, Mansuriya, Hayer, Slam Park, Arka, Dirab and King Saud University.

Fig. (7.3): Map of Riyadh region showing the position of soil samples sites, Dharme, AL-Kharj, AL-Ghad, AL-Muzahmiyya, way Qassim and Diriya

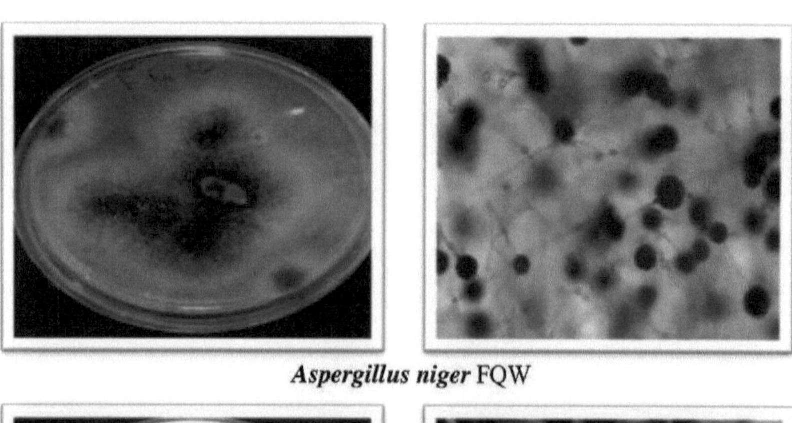

Aspergillus niger FQW

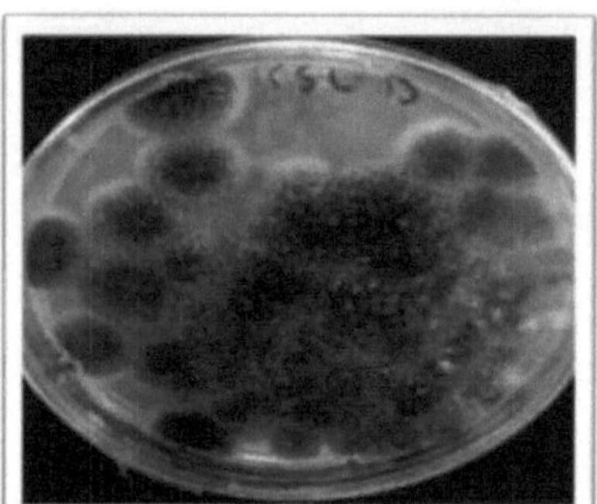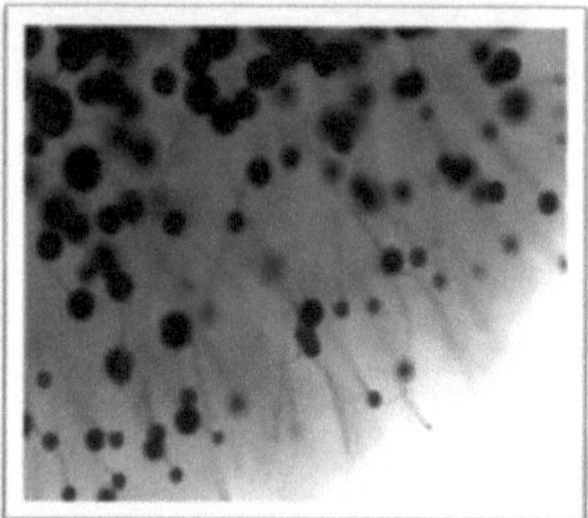

Aspergillus niger KSUD

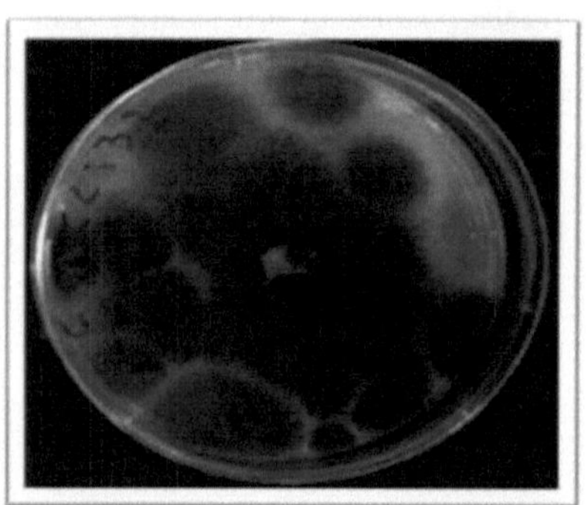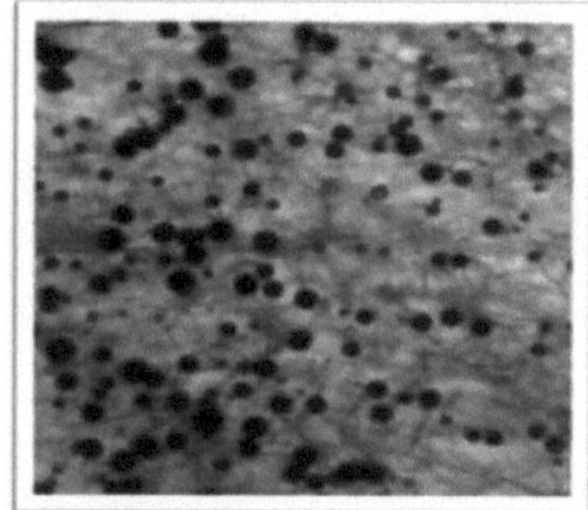

Aspergillus niger EMCC132

Fig. (7.4): Colonial characterization of *Aspergillus niger* strains

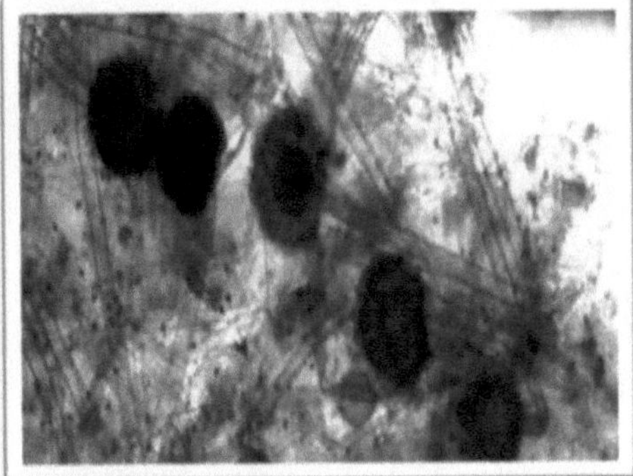

Aspergillus niger FQW

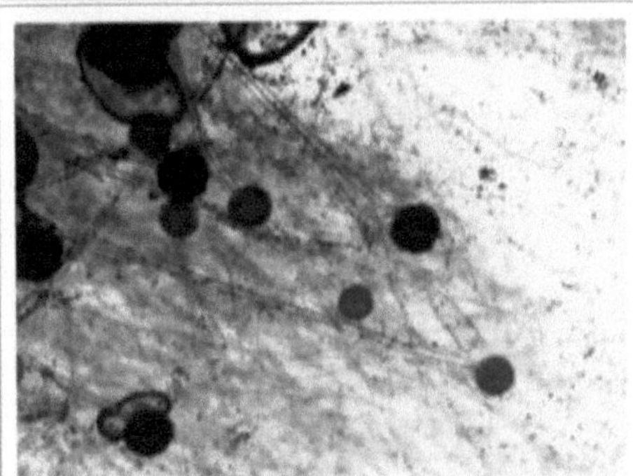

Aspergillus niger KSUD

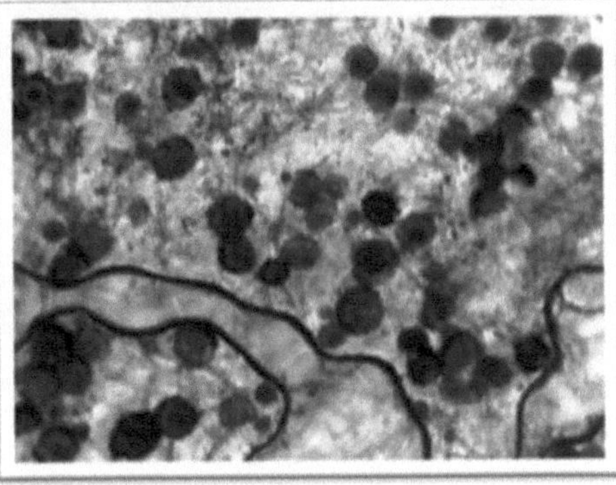

Aspergillus niger EMCC132

Fig. (7.5): Morphological characterization of *Aspergillus niger* in microscopic view.

Table (7.1): Chemical analysis of the used agricultural raw materials.

Wastes	Carbon	Hydrogen	Nitrogen	Sulfur	Phosphorus	Potassium	Calcium	Magnesium
	%				mg/kg			
Dates molasses	40.14	6.81	0.33	0.91	1204	2601	666	850
Pineapple peels	42.59	6.09	1.01	0.84	2391	5258	1567	922
Potato peels	40.95	6.61	1.20	0.85	2469	5875	361	980
Sugar Cane Bagasse	44.12	5.98	0.70	0.75	391	2752	379	592
Sugar Cane Molasses	40.93	5.94	0.20	0.80	Not Detected	3123	1710	411

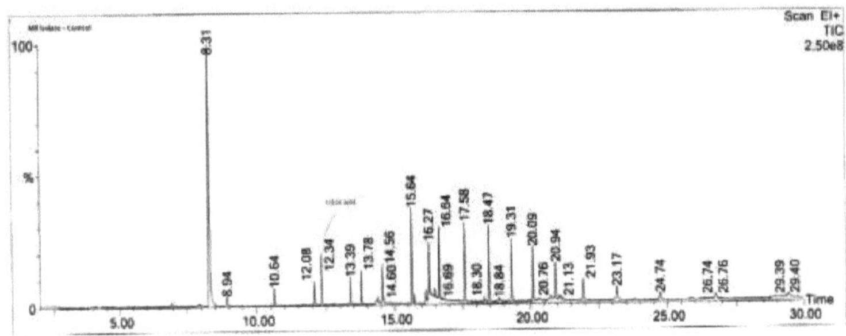

Fig. (7.6): HPLC analysis sheet of citric acid produced by untreated spores (control) of *Aspergillus niger* EMCC132.

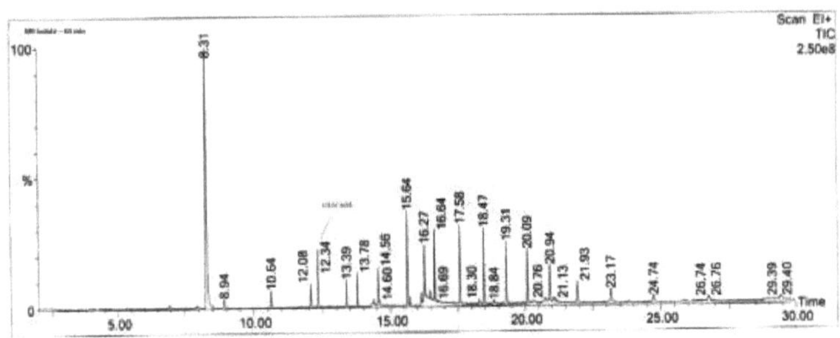

Fig. (7.7): HPLC analysis sheet of citric acid produced by spores of *Aspergillus niger* EMCC132 exposed to UV for 60 min.

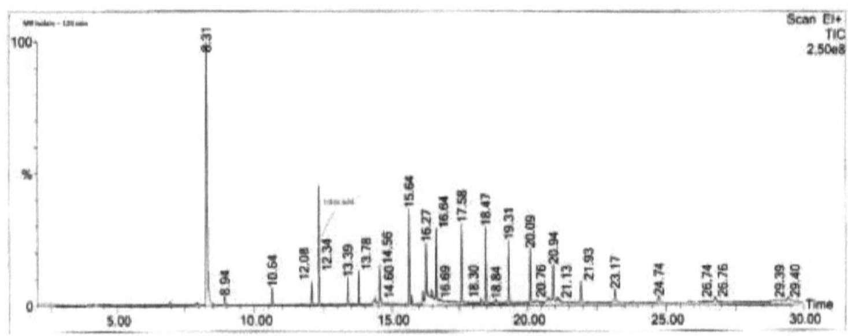

Fig. (7.8): HPLC analysis sheet of citric acid produced by spores of *Aspergillus niger* EMCC132 exposed to UV for 120 min.

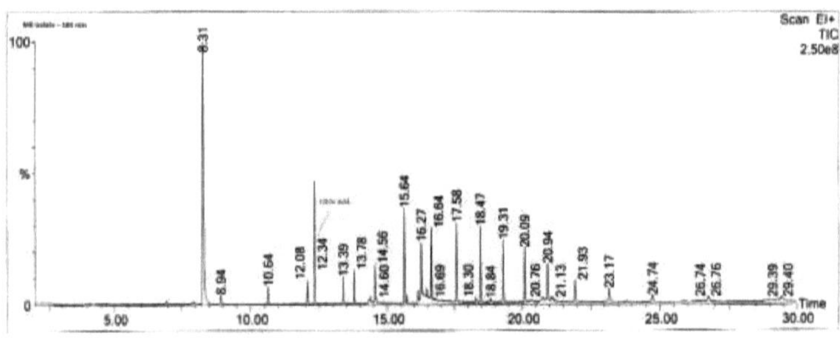

Fig. (7.9): HPLC analysis sheet of citric acid produced by spores of *Aspergillus niger* EMCC132 exposed to UV for 180 min.

الملخص العربي

يعتبر حمض السيتريك من أهم وأكثر المنتجات الطبيعية التي يستمر الطلب عليها والاحتياج لها في الاسواق العالمية. على الرغم من أن العمليات التخميرية المستخدمة لإنتاج حمض السيتريك من العمليات التخميرية المعروفة منذ زمن بعيد ، ولكن لايزال الاعتماد عليها والعمل بها يتزايد بمرور الزمن ، لذا يهدف هذا البحث الى وفرة وزيادة إنتاجية حمض السيتريك محلياً بإستخدام سلالات فطرية محلية.

يتحقق هدف تلك الدراسة من خلال النقاط التالية: اولاً عزل سلالات محلية من فطر الاسبرجيلس نيجر *Aspergillus niger* وحصر العزلات الاكثر كفاءة في إنتاح حمض السيتريك ، ثانياً تحديد السلالات الأكثر إنتاجية في أقصر فترة ممكنة ، ثالثاً دراسة بعض العوامل البيئية والغذائية الملائمة لنمو السلالات الفطرية وتحسين إنتاجية حمض السيتريك ، رابعاً دراسة إمكانية إستخدام المخلفات الزراعية لإنتاج حمض السيتريك ، خامساً دراسة إمكانية تأثير المواد المطفرة (الفيزيائية و الكيماوية) على أفضل سلالات الفطر لتحسين انتاجية حمض السيتريك.

ملخص لأهم النتائج المتحصل عليها :

1- أمكن الحصول على 96 عزلة فطرية من 20 عينة تربة تم جمعها من منطقة الرياض بالمملكة ، وطبقاً للخصائص المزرعية والميكروسكوبية تم تحديد وتعريف جميع العزلات على أنها اسبرجيلس نيجر *Aspergillus niger* .

2- وبإستخدام تقنية الاطباق الكاشفة Indicator plates ، وجد أن جميع عزلات الاسبرجيلس نيجر كانت ذات قدرة على إنتاج حمض السيتريك ولكن بكميات متباينة ، أمكن تحديد ستة عزلات للاسبرجيلس نيجر ذات أعلى قدرة إنتاجية لحمض السيتريك بعد 24 ساعة تحضين.

3- وبإستخدام تقنية مزارع الدوارق المهتزة لتنمية الفطريات والكشف الكمي عن حمض السيتريك المنتج بعد 3 أيام تحضين ، تم الحصول على أعلي إنتاجية من كل من العزلتين المحليتين Aspergillus niger FQW and KSUD ، وتم إختيارهم في التجارب اللاحقة مقارنة بالعزلة المرجعية A. niger EMCC132 والمعروفة بقدرتها الإنتاجية لحمض السيتريك.

4- وجد أن 5 أيام تحضين للعزلات الفطرية المحلية A. niger FQW and KSUD ، هي أفضل فترة تحضين للحصول على أعلى إنتاجية معنوية من حمض السيتريك.

5- ومن خلال سلسلة من التجارب صُممت لتحديد افضل الظروف البيئية والغذائية (واحدة تلو الاخرى) للحصول على أعلى إنتاجية معنوية لحمض السيتريك من العزلات الفطرية المحلية المختارة ، كانت أفضل الظروف هي: أنسب درجة حرارة هي 28°م ، أنسب درجة حموضة هي 6.5 ، أنسب كثافة للكونيدات في اللقاح المستخدم هو 3.0 X 10^6 جرثومة لكل ملليليتر ، وأوضحت النتائج أن كمية حمض السيتريك الناتجة من المزرعة المهتزة أكثر من المزرعة الثابتة ، وكان أفضل مصدر للكربون هو سكر المالتوز وبتركيز 40 جرام لكل ليتر ، وأنسب مصدر للنيتروجين هو النيتروجين العضوي الببتون وبتركيز 3 جرام لكل ليتر ، وكانت فوسفات الصوديوم هي مصدر الفوسفات الأفضل ، علاوة على تعزيز إنتاجية حمض السيتريك بإضافة كحول الميثانول بنسبة 2 %.

6- من خلال تتابع تلك العوامل القياسية والتي ادت الى تعزيزوتحسين إنتاجية حمض السيترك ، حيث بلغت 133.31 & 109.29 ملليجرام / 10 ملليليتر وهو مايعادل من 12 الى 14 مرة زيادة إذا ما قورنت بالإنتاجية الاصلية 9.07 و 9.98 ملليجرام / 10 ملليليتر للعزلات Aspergillus niger FQW و Aspergillus niger KSUD ، على الترتيب.

7- بإستخدام المخلفات والنواتج الثانوية الزراعية المحلية (قشور الأناناس الجافة ومصاصة قصب السكر الجافة وقشور البطاطس الجافة ومولاس قصب السكر ودبس التمر) كمصدر وحيد للكربون فقط أو كمصدر لكل من الكربون والنيتروجين في الظروف البيئة والغذائية القياسية ، لإختبار قدرة السلالات الفطرية تحت الدراسة على النمو وإنتاج حمض السيتريك.

8- تم الحصول على أعلى إنتاجية معنوية لحمض السيتريك وصلت الى 256.94 ملليجرام / 10 ملليليتر بواسطة السلالة المحلية *A. niger* FQW عند استخدامها مولاس قصب السكر كمصدر وحيد للكربون .

9- بالإضافة إلى ذلك ، أنتجت السلالة المحلية *A. niger* KSUD أعلى كمية معنوية من حمض السيتريك بلغت 223.77 ملليجرام / 10 ملليليتر عند تنميتها على المخلف النباتي قشور الاناناس كمصدر وحيد للكربون والنيتروجين في بيئة النمو.

10- من خلال تقييم النتائج المتحصل عليها وخاصة إنتاج حمص السيتريك بإستخدام تلك المخلفات النباتية رخيصة الثمن والمتوفرة محلياً ، ثبت معنوياً إمكانية إستخدام المولاس (سواء من قصب السكر أو التمر) وقشور الاناناس لإنتاج حمض السيتريك ذو القيمة المادية والتجارية العالية ، مما يؤكد أهمية وضرورة إستخدام تلك المخلفات من الناحية الاقتصادية لعملية الإنتاج ومن الناحية البيئية للمحافظة على سلامة البيئة من التلوث بتلك المخلفات.

11- في محاولة لتحسين إنتاجية حمض السيتريك من خلال الحصول على طفرات من سلالات *Aspergillus niger* FQW and KSUD المحلية بإستخدام المطفرات الفيزيائية او الكيماوية ، أظهرت النتائج المتحصل عليها أن الطفرة الناتجة من السلالة الفطرية *A. niger* FQW (بعد تعرضها لأشعة UV ذات الطول الموجي 254 نانوميتر لمدة 180 دقيقة) أنتجت

أعلى كمية من حمض السيتريك 41.98 مليجرام / ليتر ، مقارنة بإنتاجية السلالة الام والتي أعطت 13.918 مليجرام / ليتر.

12- كانت طفرة السلالة *A. niger* FQW أكثر ثباتا عند تعرضها للمطفر الكيماوي MTG بمعدل 750 ملليجرام لكل ليتر لمدة 60 دقيقية ، حيث أعطت أكثر من ضعف انتاجية حمض السيتريك (31.3 ملليجرام لكل ليتر) مقارنة بإنتاجية العزلة الام (13.9 ميلليجرام لكل ليتر).

Buy your books fast and straightforward online - at one of world's fastest growing online book stores! Environmentally sound due to Print-on-Demand technologies.

Buy your books online at
www.morebooks.shop

Kaufen Sie Ihre Bücher schnell und unkompliziert online – auf einer der am schnellsten wachsenden Buchhandelsplattformen weltweit! Dank Print-On-Demand umwelt- und ressourcenschonend produziert.

Bücher schneller online kaufen
www.morebooks.shop

KS OmniScriptum Publishing
Brivibas gatve 197
LV-1039 Riga, Latvia
Telefax: +371 686 204 55

info@omniscriptum.com
www.omniscriptum.com

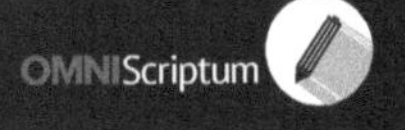

Printed by Books on Demand GmbH, Norderstedt / Germany